High-Entropy Alloys

High-Entropy Alloys

B.S. Murty
Department of Metallurgical and Materials Engineering
Indian Institute of Technology Madras, Chennai, India

J.W. Yeh
Department of Materials Science and Engineering
National Tsing Hua University, Hsinchu, Taiwan

S. Ranganathan
Department of Materials Engineering
Indian Institute of Science, Bangalore, India

AMSTERDAM • BOSTON • HEIDELBERG • LONDON
NEW YORK • OXFORD • PARIS • SAN DIEGO
SAN FRANCISCO • SINGAPORE • SYDNEY • TOKYO

Butterworth-Heinemann is an imprint of Elsevier

Butterworth-Heinemann is an imprint of Elsevier
32 Jamestown Road, London NW1 7BY, UK
The Boulevard, Langford Lane, Kidlington, Oxford, OX5 1GB, UK
Radarweg 29, PO Box 211, 1000 AE Amsterdam, The Netherlands
225 Wyman Street, Waltham, MA 02451, USA
525 B Street, Suite 1900, San Diego, CA 92101-4495, USA

Notices
Knowledge and best practice in this field are constantly changing. As new research and
experience broaden our understanding, changes in research methods, professional practices,
or medical treatment may become necessary.

Practitioners and researchers must always rely on their own experience and knowledge in
evaluating and using any information, methods, compounds, or experiments described herein.
In using such information or methods they should be mindful of their own safety and the safety
of others, including parties for whom they have a professional responsibility.

To the fullest extent of the law, neither the Publisher nor the authors, contributors, or editors,
assume any liability for any injury and/or damage to persons or property as a matter of products
liability, negligence or otherwise, or from any use or operation of any methods, products,
instructions, or ideas contained in the material herein.

British Library Cataloguing in Publication Data
A catalogue record for this book is available from the British Library

Library of Congress Cataloging-in-Publication Data
A catalog record for this book is available from the Library of Congress

ISBN: 978-0-12-800251-3

For information on all Butterworth-Heinemann publications
visit our website at **store.elsevier.com**

This book has been manufactured using Print On Demand technology. Each copy is produced
to order and is limited to black ink. The online version of this book will show color figures
where appropriate.

Working together
to grow libraries in
developing countries

www.elsevier.com • www.bookaid.org

CONTENTS

FOREWORD

In the 1970s, I became excited about the idea of multicomponent alloys. I realized that the materials we use are almost all based on a single component with a primary property and an admixture of small amounts of other components to provide secondary properties. Effectively, all our known materials are at the corners and edges of a multicomponent phase diagram consisting of all possible components. This means that there is a vast number of possible materials in the middle of this phase diagram which have never been investigated. It turns out that this unknown field of materials is truly enormous, and the number of unexplored materials is many times greater than the number of atoms in the universe.

This explains why we keep discovering exciting new materials: high-temperature superconductors, glassy alloys, quasicrystals, compound semi-conductors, and so on. And we will keep finding exciting new materials as long as we have the courage as experimenters to try innovative, new mixtures of constituents. I now tell every PhD student in materials science to be aggressive and ambitious in exploring this amazing array of potential new materials.

In the 1970s, I found it hard to persuade other people to be similarly enthused by my ideas on the topic of multicomponent alloys. I could not get funding, and I could not get research colleagues to undertake preliminary experiments. Everyone wanted to work in much better known fields. Everyone was very conservative. Finally, I persuaded a young undergraduate student, Alan Vincent, to do preliminary work. He immediately found exciting results including the first high-entropy alloys. Nearly 20 years later, I persuaded another young undergraduate student, Peter Knight, to repeat the work. And finally, a couple of years later, my long-standing research colleague, Isaac Chang, repeated the experiments for a third time, but more carefully and with more time to document the results fully and in a publishable way. We presented the results, first found by Alan Vincent in 1979, at a conference in 2002 which was published in 2004, more than 25 years since my first idea. In parallel, Professor S. Ranganathan (my old friend, Rangu) and Professor J.W. Yeh (my new friend, Jien-Wei) published independently papers on, respectively, material cocktails and high-entropy alloys, closely related and essentially similar concepts to my idea of multicomponent alloys.

In the last few years, as a consequence of the outstanding continuing work by Jien-Wei, the field of multicomponent and high-entropy alloys has taken off, with literally hundreds of publications each year. Most notably, Vincent, Knight, Chang, and I discovered in the late 1970s a single FCC solid solution consisting of five components in equal proportions, namely, FeCrMnNiCo. This alloy has been shown to have outstanding mechanical properties, with high strength and high ductility. I realized in the late 1970s that the mechanical behavior of this material would be very unusual. Metal and alloy mechanical properties depend primarily on the behavior of dislocations and how they move in response to stress, but the concept of a dislocation as a line defect with a consistent core structure becomes complex when there are many different components distributed on a single lattice.

Professors Murty, Yeh, and Ranganathan have now written a book on this new group of materials. The book covers the structure, processing, and properties of the materials, insofar as we have been able to explore them. Some multicomponent alloys are solid solutions with high entropy. Some are not. In either case, there are wonderfully exciting new structures and properties to be found. This book is the first about this field. It contains many valuable and interesting insights. But ultimately it can only hint at the full range of new materials which remain to be discovered. The authors should be congratulated on doing an important job which will help us on our exciting, exploratory journey into the materials of the future.

Prof. Brian Cantor
Vice Chancellor,
The University of Bradford, UK

PREFACE

Alloys traditionally have been based on a solvent element to which various solute atoms are added for improving specific properties. Thus, alloys are usually named after the major element in the alloy (e.g., Fe-, Al-, Cu-, Mg-, and Ni-base alloys). Two people, in recent times, have changed the way people look at alloys and they are Prof. Brian Cantor and Prof. Jien-Wei Yeh by coming up with equiatomic and nonequiatomic multicomponent alloys. Incidentally, though each of them started working on these alloys independently at different times (Cantor starting in 1979 and Yeh starting 1996), their work came to open literature in the same year, 2004. Interestingly, even before the papers of these two pioneers got published in 2004, Prof. S. Ranganathan felt the importance of this new class of alloys and wrote about them in his classic paper "Alloyed Pleasures: Multimetallic Cocktails" in 2003, which has been cited more than 100 times now as the first publication on this class of alloys.

Yeh christened these alloys as "high-entropy alloys (HEAs)," rightly so, as the configurational entropy of these alloys is expected to be very high at their random solution states. Such a high entropy is expected to drive the tendency to form simple solid solutions (crystalline or amorphous) rather than complex microstructures with many compounds. The concept has caught the attention of many researchers and the last one decade witnessed about 400−500 papers being published on HEAs with various elemental combinations. Two major observations can be made from all this work, namely, the alloys do form simple solid solutions in most of the cases and the number of phases observed in these alloys is much less than the maximum predicted from the Gibbs phase rule.

There are also clear indications that the high entropy in these multicomponent equiatomic and nonequiatomic alloys is not able to act like a glue holding all the atoms together in a single solid solution, and there are reports on the formation of two or more phases in which intermetallic phase formation and even segregation of certain elements were observed. This could be related with the various thermodynamic and kinetic factors. There has been intense activity in past few years to predict the phases that can form in such multicomponent alloys through various modeling approaches including integrated computational materials engineering (ICME) using various tools such as CALPHAD, *ab-initio*, molecular dynamics, Monte Carlo, and phase field approaches, which have been supported by materials genome initiative (MGI).

Besides the scientific curiosity, researchers also feel that HEAs can substitute conventional materials in advanced applications so that the limitations of the latter in service life and operational conditions could be overcome by providing superior performance of the former. A number of processing routes, including conventional melting and casting, mechanical alloying, various coating techniques, and even combinatorial materials science approaches are being used to synthesize and process this new class of alloys. There have been a lot of studies on understanding both the structural and functional properties of these alloys. The results of HEAs and HEA-related materials reported so far by various research groups are very encouraging for their applications in a wide range of fields such as materials for engine, nuclear plant, chemical plant, marine structure, tool, mold, hard facing, and functional coatings.

It is just over a hundred years since Walter Rosenhain wrote his seminal book on Physical Metallurgy. A century of research resulted in spectacular progress but the field became mature and the excitement began to wane. At the beginning of the third millennium, the discovery of High-Entropy Alloys has ushered in a renaissance in physical metallurgy.

This book presents a comprehensive insight into all the above aspects of this exciting new class of alloys. The book, being a short format one, is written keeping a beginner in the field in mind to give him/her an idea on various facets of HEAs that he/she can pursue.

B.S. Murty, J.W. Yeh and S. Ranganathan

ACKNOWLEDGEMENTS

The authors are grateful to Prof. Brian Cantor for readily agreeing to write a foreword for the book, in spite of his hectic schedule as Vice Chancellor of the University of Bradford, UK. They are also highly indebted to all the authors of various papers that are referred to in this book. Their contributions have enhanced the quality of the book.

Prof. Murty would like to specially thank his group members, Mayur, Guruvidyathri, Anirudh, Arul, Ameey and Dr. Sanjay, who have untiringly helped at various stages of the book. He would also like to gratefully acknowledge the collaboration with his students, Varalakshmi, Praveen, Pradeep, Sriharitha, Ashok, Durga, Raghavan and collaborators Dr. Ravi Sankar Kottada, Prof. M. Kamaraj and Prof. K.C. Hari Kumar, Dr. Sheela Singh, Dr. N. Wanderka and Prof. J. Banhart and Prof. D. Raabe. He is also indebted to his family members for their patience and continuous support.

Prof. Yeh would like to thank two important group members, Prof. Su-Jien Lin and Prof. Swe-Kai Chen, for their long-term contributions in the research on high-entropy alloys since 1995. He also expresses sincere thanks to Profs. Tsung-Shune Ch in, Jan-Yiaw Gan, Tao-Tsung Shun, Chun-Huei Tsau, Shou-Yi Chang, Tung Hsu, Wen-Kuang Hsu and all graduate students for their significant efforts and contributions in HEA-related research. In addition, he specially thanks Dr. Chun-Ming Lin and PhD student Chien-Chang Juan for their help in searching related papers, plotting figures and providing suggestions for this book. Finally, he gratefully thanks his beloved wife and daughters for all their encouragement and support during writing this book.

Prof. Ranganathan records his gratitude to numerous teachers, colleagues and students who instilled in him an abiding interest in physical metallurgy. Special thanks are due to T.R. Anantharaman, K.P. Abraham, C.N.R. Rao, P. Rama Rao, Alan Cottrell, David Brandon, Robert Cahn, A.L. Mackay, Gareth Thomas, J.W. Cahn, P. Ramachandra Rao, K. Chattopadhyay, D. Shechtman, K.H. Kuo and A. Inoue. Discussions with T.A. Abinandanan and Abhik Choudhury were stimulating. The work of Sudarshna Kalyanaraman on HEA is also acknowledged with appreciation. Grateful thanks are expressed to members of his family for support in writing this book.

The authors are deeply indebted to the meticulous reading of the entire book by Dr. R. Krishnan and thank him profusely for his useful suggestions.

B.S. Murty, J.W. Yeh and S. Ranganathan

CHAPTER 1

A Brief History of Alloys and the Birth of High-Entropy Alloys

1.1 INTRODUCTION

Alloying is the greatest gift of metallurgy to humankind. The English language insists on unalloyed pleasures, thereby implying that the sensation of pleasure must be pure and not admixed with other emotions. Exactly the opposite rules in metallurgy, where pure metals have few uses but lot more upon alloying. The power of this idea of alloying is not confined to metals. The same principle of alloying applies in polymers and ceramics. It can be carried further by mixing two classes of materials to create a variety of composites.

The civilizational journey of humankind began with the discovery of native metals such as gold and copper as pure metals. Nowadays we have access to an incredible number and variety of materials. Ashby map (Ashby, 2011) shown in Figure 1.1 gives a panoramic view of the development in the use of materials over 10 millennia. A graphic depiction of the different classes of materials from ceramics to metals, polymers, and more recently to composites is vividly displayed. The passage from discovery through development to design of materials can be noted. Ashby's (2011) map in term of strength versus density shown in Figure 1.2 demonstrates the filling of material–property space in a vivid fashion from 50,000 BCE up to the present scenario. In time scale, the largest filling has occurred in the past 50 years during which envelopes of metals, ceramics, and composites had a large expansion, and new envelopes of synthetic polymers and foam materials take a significant space. But, the filled area also seems to approach some fundamental limits beyond which it is difficult to go further (Ashby, 2011).

In many ways, the history of alloying is the history of metallurgy and materials science. Books and treatises have been written. An elegant and brief history is by Ashby (2008). Cahn (2001) has offered a magisterial survey of "The coming of materials science". Ranganathan (2003) wrote on alloyed pleasures—an ode to alloying. In the following sections, a few episodes in this epic journey are covered.

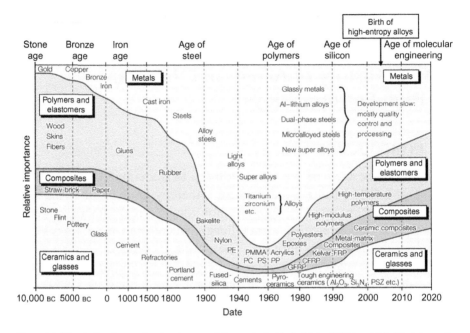

Figure 1.1 Historical evolution of engineering materials—marked with the birth of HEAs published in Advanced Engineering Materials (Yeh et al., 2004b). Adapted from Ashby (2011).

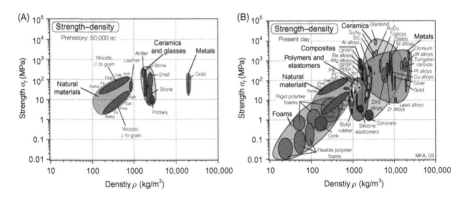

Figure 1.2 The explosion in the diversity of materials in the modern era (Ashby, 2011); (A) prehistoric era (50,000 BCE) and (B) current status.

1.2 THE COMING OF ALLOYS

Native alloys such as tumbaga and electrum are alloys of gold−copper and gold−silver, respectively. When platinum was discovered in 1735, it was compared with silver. Also, mixtures of platinum metals are found to occur in nature. It is an early example of multicomponent high-entropy alloys (HEAs), since platinum is often found as alloys with the other platinum group metals and iron mostly.

Alloying was an accidental discovery. In the primitive fires in the caves, ores of copper got mixed with ores of arsenic, zinc, and tin. The first alloy of copper and arsenic (Arsenical bronze, 3000 BCE) was entirely accidental. A more intentional alloying of tin with copper (tin bronzes in 2500 BCE) gave birth to the Bronze Age, as bronze was superior in its mechanical properties.

The seven metals found in antiquity were gold, copper, silver, iron, lead, tin, and mercury. The eighth metal, zinc, is added because of the unique Indian context but also because the discovery of other metals had to await the advent of the scientific revolution for a few centuries.

It is interesting to mention that intermetallics of copper—tin alloys had been used in ancient time. Mirrors were made of bronzes in different parts of the Old World including India and China, due to their higher hardness which makes it easy in getting mirror finish to reflect like silver. Archaeo-metallurgical investigations by Sharada Srinivasan on vessels from South Indian megaliths of the Nilgiris and Adichanallur (1000—500 BCE) showed that they were of wrought and quenched high-tin beta bronze, ranking among the earliest known artifacts. This is an early application of an intermetallic. When the sulfide ores of copper and nickel were smelted together, it led to copper—nickel alloy in the fourth century in China. Zinc was added in the twelfth century to form silvery and rust-resisting alloy known as paktong (white copper), which was widely used in Europe before stainless steel was invented.

Wrought iron was produced as early as in 1000 BCE and cast iron and cast steel were produced one millennium later. Steel was an accidental alloy of iron with carbon. This is all the more astonishing as carbon was not recognized as an element until recently. This accidental discovery also led to the production of wootz steel in India around 300 BCE. This has been rightly celebrated as the most advanced material of the ancient world, as this steel was used to fashion the Damascus swords. The deciphering of wootz steel by European scientists led to the correlation between structure and properties at first and subsequently between composition, processing, structure, and properties. Figure 1.3 shows that materials hypertetrahedron that links

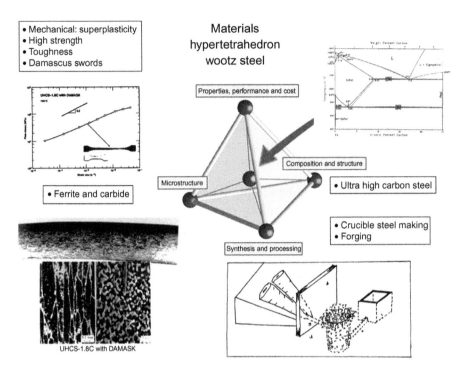

Figure 1.3 The materials hypertetrahedron for Wootz Steel (Srinivasan and Ranganathan, 2014).

the above four with modeling is necessary to understand the performance of wootz steel. (Ranganathan and Srinivasan, 2006; Srinivasan and Ranganathan, 2014). The facets of the ultrahigh carbon steels, Buchanan furnace, the Fe–C phase diagram, the microstructure of dendrites in the as-cast state and spheroidized cementite in the forged state, the superplastic elongation, and the Damascene marks are emphasized for the strong interconnections among them. It can be regarded as a classic example of the materials tetrahedron but including a fifth vertex of modeling (e.g., CALPHAD method to calculate phase diagram) to make it a hypertetrahedron.

When the first industrial revolution began in later half of 18th century in England, more and more elements were found and produced by humankind. From these "new" elements, numerous metallic materials including engineering and advanced alloys have been developed. They were synthesized with different compositions and produced by various processing routes. Up to now, about 30 alloy systems, each system based on one principal metallic element, have been developed and used for a variety of products (Handbook Committee, 1990).

Several alloys of engineering importance have been developed. Michael Faraday created some of the first alloy steels in his efforts to reproduce the wootz steel from India and is hailed as the father of alloy steel. Aluminum alloys were produced after the commercialization of Hall-Héroult process in the mid 1880s and underwent large progress in precipitation hardenable alloys such as Al–Cu–(Mg) and Al–Zn–Mg–(Cu) for light and strong airframes in the explosive expansion of the airplane industry during and after World War I (1914–1918). High-speed steel for cutting tools was first produced in the early 1900s. To meet the challenge for even higher cutting speed, cemented or sintered carbides of WC/Co composites were introduced in the 1930s. At the same time, superalloy development began in the United States in the 1930s and was accelerated by the demands of gas turbine technology. Ferritic, austenitic, and martensitic stainless steels were almost simultaneously developed around 1910.

1.3 SPECIAL ALLOYS

Besides the above modern engineering alloys, several special alloy systems with specific compositions, structures and properties have been developed with intensive research in last 50 years. They are intermetallics, quasicrystals, and metallic glasses as introduced in the following sections.

1.3.1 Intermetallics and Quasicrystals

Intermetallic compounds are basically compounds of two or more metallic elements. They are brittle by nature. Besides the ancient intermetallic mirrors made of high-tin bronze mentioned in last section, they have given rise to various novel materials developments in modern time including magnetic AlNiCo alloys and the $LaNi_5$ for nickel metal hydride batteries, and various aluminides, Ni–Al-, Ti–Al-, Fe–Al-based compounds for elevated-temperature lightweight applications in turbine engines.

Another class of intermetallics includes those which demonstrate so called forbidden rotational symmetries (such as 5- and 10-fold rotational symmetries) and quasiperiodic translational symmetry. These were discovered by Shechtman (1984) in 1982 as described in his notebook, when he first observed a 10-fold electron diffraction pattern from a rapidly solidified Al–Mn alloy. These were christened as quasicrystals. This observation has shaken the beliefs of crystallographers to the extent that the definition of a crystal has been modified in 1990, based on the discovery of this new class of materials. There has been intensive research on these exciting materials in the past three decades, both towards the understanding of the structure and properties of these materials.

1.3.2 Metallic Glasses

The first reported metallic glass was an alloy ($Au_{75}Si_{25}$) produced at California Institute of Technology by the research group of Pol Duwez (Klement et al., 1960), which was cooled rapidly from the liquid state at a rate of around 10^6 K/s to avoid crystallization and possess noncrystalline or glass-like structure. Among the different metallic glasses developed later, the soft magnetic metallic glass of iron, nickel, phosphorus, and boron, known as Metglas, was commercialized in early 1980s and is used for low-loss power distribution transformers. In 1974, H.S. Chen first reported that bulk metallic glass rods in diameters ranging from 1–3 mm can be produced in various glassy ternary alloys including Pd–Au–Si, Pd–Ag–Si, and Pt–Ni–P systems (Chen, 1974). After this discovery, multicomponent glassy alloys based on lanthanum, magnesium, zirconium, palladium, iron, copper, and titanium, etc. with critical cooling rate in the range of 1–100 K/s, comparable to oxide glasses have been developed and researched.

1.4 THE COMING OF MULTICOMPONENT HEAs

From the above description of conventional and special alloys, historically over five millennia the alloy design, alloy production, and alloy selection were all based on one principal-element or one-compound concept. This alloy concept has generated numerous practical alloys contributing to civilization and daily life. But, it still limits the degree of freedom in the composition of the alloy and thus restricts the development of special microstructures, properties, and applications. Consequently, materials science and engineering of alloys is not fully explored since those alloys outside this conventional scheme have not been included.

1.4.1 Karl Franz Achard

It should be mentioned that in the late of eighteenth century, a German scientist and also metallurgist Franz Karl Achard had studied the multicomponent equimass alloys with five to seven elements (Smith, 1963). He could be most probably the first one to study multiprincipal-element alloys with five to seven elements. In many ways, he is the predecessor for the researches of Jien-Wei Yeh on HEAs. More than two centuries separate them. In 1788, Achard published a little-known French book "Recherches sur les Propriétés des Alliages Métallique," the first compilation of data on alloy systems in Berlin. He disclosed the results of a laborious and comprehensive program on over 900 alloy compositions of 11 metals, including iron, copper, tin, lead, zinc, bismuth, antimony, arsenic, silver, cobalt, and platinum. Because of high cost, he studied fewer compositions with silver, cobalt, and platinum. For other elements, he made representative alloys of every possible combination of components up to seven components. Besides many binary, ternary, and quaternary alloys, he made quinary, sexinary, and septenary alloys only in equal proportions in weight. All the alloys were in the as-cast condition and on these he carried out tests for density, hardness, strength, impact resistance, ductility, the resistance to a file, the degree to which the alloy could be polished, and finally the results of exposing a polished surface to dry air, moist air, and moist air with HCl acid fumes, and moist hydrogen sulfide. In this book, he pointed out that the properties of alloys are quite different from those of the pure metals and are unpredictable. Only experiment can instruct us. This book is mainly a report without any discussion. All the experimental results were given in tabular form. Although this book was written in the French language insisted upon by Frederick the Great, but unfortunately it was virtually ignored by metallurgists everywhere. This rare work was brought to light only in 1963 by Professor Cyril Stanley Smith (Smith, 1963).

Toward the end of the twentieth century two entirely independent investigations by Brian Cantor in the United Kingdom and Jien-Wei Yeh in Taiwan made a disruptive break with the classical tradition of alloys. A brand new alloy concept "HEAs" has been proposed and explored and has led to a flurry of excitement. Figure 1.4 gives the number of year-wise journal publications (until 2013) in the area of HEAs.

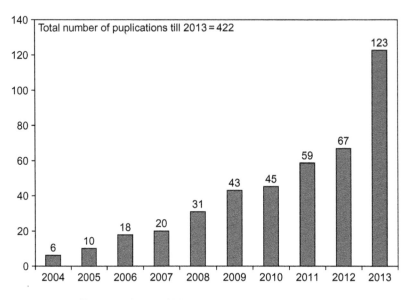

Figure 1.4 Year-wise publications in the area of HEAs.

1.4.2 Brian Cantor

The first work on exploring this brave new world was done in 1981 by Cantor with his student Alain Vincent. They made several equiatomic alloys mixing many different components in equal proportions. In particular, the world record holding multicomponent alloy consisting of 20 different components each at 5% is held by this study. It was noticed that only one alloy with a composition of $Fe_{20}Cr_{20}Ni_{20}Mn_{20}Co_{20}$ forms a single FCC (face centred cubic), Vincent was an undergraduate project student and the work was only written at that time in his thesis at Sussex University.

After this initial experiment there was a hiatus. Similar studies on a wider range of alloys were repeated with another undergraduate project student, Peter Knight, at Oxford in 1998. He achieved some similar results and some new ones, published his results in a thesis at Oxford. Finally, Isaac Chang repeated the work again in about 2000 at Oxford, and finally published the results in the open literature by presenting at the Rapidly Quenched Metals conference in Bangalore in 2002, which was then published in the journal *Material Science and Engineering A* in July 2004 (Cantor et al., 2004). In this paper entitled "Microstructural development in equiatomic multicomponent alloys", several important conclusions were drawn. A five component $Fe_{20}Cr_{20}Mn_{20}Ni_{20}Co_{20}$ alloy, on melt spinning, forms a single FCC solid solution, which solidifies dendritically. Based on this composition, a wide range of six to nine component alloys by adding other elements such as Cu, Ti, Nb, V, W, Mo, Ta, and Ge exhibit the same majority FCC primary dendritic phase, which can dissolve substantial amounts of other transition metals such as Nb, Ti, and V. More electronegative elements such as Cu and Ge are less stable in the FCC dendrites and are rejected into the interdendritic regions. Besides, alloy containing 20 components, that is, 5 at.% each of Mn, Cr, Fe, Co, Ni, Cu, Ag, W, Mo, Nb, Al, Cd, Sn, Pb, Bi, Zn, Ge, Si, Sb, Mg, and another alloy consisted of 16 elements, that is, 6.25 at.% each of Mn, Cr, Fe, Co, Ni, Cu, Ag, W, Mo, Nb, Al, Cd, Sn, Pb, Zn, and Mg are multiphase, crystalline and brittle both in as-cast condition and after melt spinning. Surprisingly, however, the alloys consisted predominantly of a single FCC primary phase, containing many elements but particularly rich in transition metals, notably Cr, Mn, Fe, Co, and Ni. Finally, the total number of phases is always well below the maximum equilibrium number allowed by the Gibbs phase rule, and even further below the maximum number allowed under non-equilibrium solidification conditions.

It is also important to point out that Cantor came up with another novel idea of equiatomic substitution later, in the early 2000s (Kim et al., 2003a), as a method of exploring metallic glass. These compositions are also in this vast uncharted region of materials space.

1.4.3 Jien-Wei Yeh

J.W. Yeh independently explored the multicomponent alloys world since 1995 (Yeh et al., 2004b; Hsu et al., 2004). Based on his own concept that high mixing entropy factor would play an important effect in reducing the number of phases in such high order alloys and render valuable properties, he supervised a master student K.H. Huang in 1996 to start the research and see the possibility of success in the fabrication and analysis of HEAs. Around 40 equiatomic alloys with five to nine components were prepared by arc melting. Investigations were made on microstructure, hardness, and corrosion resistance of as-cast state and fully annealed state. The alloy design is mainly based on commonly used elements. From those data of around 40 compositions, 20 alloys based on Ti, V, Cr, Fe, Co, Ni, Cu, Mo, Zr, Pd, and Al, with or without 3 at.% B addition were selected as experimental alloys in the MS thesis of Huang in 1996 (Huang, 1996, published as MS thesis of National Tsing Hua Univeristy, Taiwan).

Based on this study, typical dendritic structure was seen in the as-cast structure. All alloys have high hardness level in the range from 590 to 890 HV depending on the composition and fabrication conditions: as-cast or fully annealed. Full annealing treatment in general retained similar hardness level of as-cast state. Higher number of elements increased the hardness but nine-element alloys more or less displayed a small decrease in hardness. Small addition of B has led to some increase in hardness.

After this study, two more studies were made before 2000 on different aspects of HEA and were submitted as MS theses (Lai et al., 1998 and Hsu et al., 2000, all published as MS theses of National Tsing Hua University, Taiwan). During 2001–2003, nine different studies were conducted by Professor Yeh's group: five studies on HEAs bulk alloy concerning with deformation behavior, wear behavior, and annealing behavior; two studies on HEA thin films deposited by magnetron sputtering; and two on HEA thermal spray coatings (Huang et al., 2001; Chen et al., 2002; Tung et al., 2002; Chen and Lin, 2003; Huang and Yeh, 2003; Hsu et al., 2003; Lin et al., 2003; Tsai et al., 2003a; Tsai et al., 2003b, all published as MS theses of National Tsing Hua University, Taiwan). Until 2013, Professor Yeh has supervised 79 MS theses and 10 Ph.D. theses in this exciting area of HEAs. Besides the supervision by Yeh, a portion of some of these theses on HEAs and related materials were also supervised by his colleagues and collaborators: S.K. Chen, S.J. Lin, T.S. Chin, J.Y. Gan, and T.T. Shun. In the discussion on these trends, high solution hardening due to large lattice distortion and stronger bonding were proposed. All these alloys in general displayed very good corrosion resistance assessed by the weight loss after immersion in four acidic solutions of HCl, H_2SO_4, HNO_3, and HF, each in 0.01 and 1 M, for 24 h. The addition of passive elements and the benefit of low free energy due to high entropy were thought to contribute the corrosion resistance. This study thus led to valuable suggestions about high-entropy effect, lattice distortion effect, and slow diffusion effect.

Yeh had submitted the "HEA concept" paper to *Science* in January 2003 but finally unaccepted by *Science*. After this, he submitted the same paper to *Advanced Materials* and then agreed the transfer to her sister journal, *Advanced Engineering Materials* for publication. In May 2004, this paper entitled "Nanostructured high-entropy alloys with multiprincipal elements—novel alloy design concepts and outcomes" was published. It becomes the first one to elucidate the concept of HEAs by providing experimental results and related theory (Yeh et al., 2004b). Besides this, another paper entitled "Multi-principal-element alloys with improved oxidation and wear resistance for thermal spray coating" was published in *Advanced Engineering Materials* in February 2004 (Huang et al., 2004). But the term of HEA was not used in this paper. Two papers entitled "Wear resistance and high-temperature compression strength of FCC CuCoNiCrAl$_{0.5}$Fe alloy with boron addition" and "Formation of simple crystal structures in solid-solution alloys with multi-principal metallic elements" were published in *Metallurgical and Materials Transactions A* later in the same year (Hsu et al., 2004; Yeh et al., 2004a). Before the submission of the first of the above paper, Professor Yeh had applied for HEAs patents in Taiwan (December 10, 1998), Japan, United States, and Mainland China.

1.4.4 Srinivasa Ranganathan

Professor S. Ranganathan has also spent a long time to look into such multicomponent alloys unexplored by people. Through the communications and discussions on this unknown field with J.W. Yeh, he published a paper entitled "Alloyed pleasures—multimetallic cocktails" to introduce three new alloy areas: bulk metallic glasses by A. Inoue, superelastic and superplastic alloys (or gum metals) by T. Saito, and HEAs by J.W. Yeh in *Current Science* in November 2003 (Ranganathan, 2003). This becomes the first open publication in journals on HEAs, which led to the activation of this new field.

In this article, he said that the multicomponent alloys represent a new frontier in metallurgy. They require hyperdimensions to visualize. If we use a coarse mesh of 10 at.% for mapping a binary system, the effort involved in experimental determination of phase diagrams rises steeply. Thus, the effort of experimental determination of a seven component system will be 10^5 times that of a binary diagram and will alone need as much effort as has been spent over the last 100 years in establishing ~ 4000 binary and ~ 8000 ternary diagrams. While the computation of phase diagrams from first principles has made impressive progress in the last decade, the calculation of higher order systems is a daunting task. In this scenario, we have explorers like A. Inoue, T. Saito and J.W. Yeh pointing to exciting new alloys with applications.

1.5 THE SCOPE OF THIS BOOK

From the open publications by the three initiators mentioned above, HEAs have become an emerging field with many more researcher's efforts and contributions. In a broad view, many aspects have been

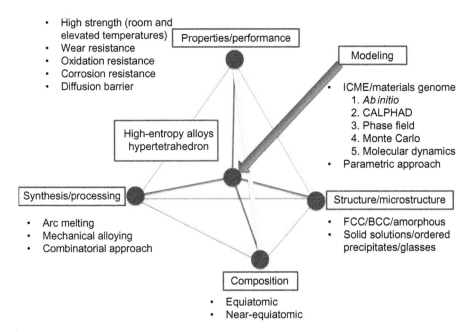

Figure 1.5 The materials hypertetrahedron for HEAs.

explored and researched. Figure 1.5 shows the materials hypertetrahedron for the HEAs, which shows in a nutshell the broad spectrum of research and development that is taking place in this field. Based on this, the content of this book has been designed to cover this broad spectrum. Chapter 2 gives the basic definition of HEA and the compositional notations. Chapter 3 describes the factors affecting the phase selections and gives the parametric approaches to design alloy compositions. Chapter 4 describes the different simulation and modeling methods involved in integrated computational materials engineering (ICME) and material genome initiative (MGI). Chapter 5 describes the different synthesis methods used for HEAs. Chapter 6 describes the formation of solid solutions and their microstructure in different HEA compositions. Chapter 7 describes the intermetallics, interstitial compounds and metallic glasses found in HEAs. Chapters 8 and 9 describe the structural and functional properties, respectively. Finally, Chapter 10 describes property goal pursued, potential advanced applications and future trends. As for the interconnections in the HEAs hypertetrahedron, all chapters also emphasize them with an aim to give comprehensive understanding on HEAs.

CHAPTER 2

High-Entropy Alloys: Basic Concepts

2.1 INTRODUCTION

As the combinations of composition and processes for producing HEAs are numerous and as each HEA has its own microstructure and properties to be identified and understood, the research work is truly limitless (Zhang et al., 2014). It becomes very important to consider the basic concepts relating to HEAs at the very beginning, including the origin of high entropy, classification, definition, composition notation, and the four core effects of HEAs.

2.2 CLASSIFICATION OF PHASE DIAGRAMS AND ALLOY SYSTEMS

Phase diagrams and the systems they describe are often classified and named after the number, in Latin, of components in the system. They can be christened based on their number of components as unary, binary, ternary, quaternary, quinary, sexinary, septenary, octonary, nonary, and denary from single- to 10-component alloys, respectively. It is suggested that this nomenclature recommended by *ASM Handbook Volume 3* (1992) may be followed. This volume acquires renewed interest as phase diagrams are the beginning of wisdom, as observed by Hume-Rothery.

There are very few binary engineering alloys that are commercially in use and most alloys are actually multicomponent alloys. However, most of these multicomponent alloys are still based on one principal element and are named after the principal elements, such as Fe-, Al-, Cu-, Ni-, and Ti-based alloys. These alloys may consist of solid solutions and/or intermetallic compounds (crystalline or quasicrystalline). These often include interstitial compounds, known as Hagg phases. Under nonequilibrium conditions some of the alloys form metallic glasses.

Cantor (2007), has recently shown that the total number of possible alloys (N) that can be formed with C number of components, when each alloy differs in composition by $x\%$, can be written as $N = (100/x)^{C-1}$.

If we consider 60 different elements (excluding those that are too radio-active, toxic, rare, and/or otherwise difficult to use) in the periodic table to combine with each other to form alloys, the total number of possible alloys turns out to be about 10^{177}, if each alloy differs in composition by 0.1%. Even if we reduce the number of elements to 40 that can be used to make the alloys and considering that each alloy differs in composition by 1%, the total number of possible alloys turns out to be 10^{78}, which is an astronomical number, considering that there are only 10^{66} atoms in the galaxy (Cantor, 2007).

The mutual solubility between solvent and solute components in a binary alloy system could be judged by Hume-Rothery rules, namely, crystal structure, atomic size difference, valence, and electronegativity. In fact, all these factors also influence the interaction between different elements and make the enthalpy of mixing either negative (attractive interaction leading to ordering and the formation of intermetallic compounds), positive (repulsive interaction leading to clustering and segregation), or near zero (leading to the formation of disordered solid solutions). The competition between enthalpy of mixing and entropy of mixing further affects the solubility between two components. When solubility is limited, terminal solid solutions based on each component can be obtained in the phase diagram. When a solid solution forms at all compositions, it is called an isomorphous system. But continuous solid solutions in binary alloy system are not common because the conditions for its formation are very strict to fulfill. Available binary alloy phase diagrams (Massalski, 2001) indicate that the total number of isomorphous binary and ternary systems is only 153 and 248, respectively.

While such terminal solutions are well known in binary, ternary and quaternary alloy systems, it is curious to know whether we can obtain solid solution phases in the center of higher order phase diagrams or not. In fact, formation of such solid solutions in the center of the phase diagram has not been explored much (Figure 2.1). In a significant deviation from traditional ways of making alloys, Cantor et al. (2004) and Yeh et al. (2004b) independently came up with the idea of preparing equiatomic or near-equiatomic multicomponent alloys. Yeh et al. (2004b) popularized these alloys as "HEAs" by pointing out the known thermodynamic fact that configurational entropy of a binary alloy ($\Delta S_{conf} = -R(X_A \ln X_A + X_B \ln X_B)$) is a maximum when the elements are in equiatomic proportions (Figure 2.2) and that the maximum

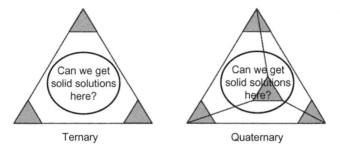

Figure 2.1 Schematic ternary and quaternary systems, showing regions of the phase diagram which are relatively well known (green) near the corners, and relatively less well known (white) near the center. (For interpretation of the references to color in this figure legend, the reader is referred to the web version of this book.) Adapted from Cantor (2007).

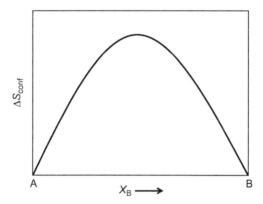

Figure 2.2 Configurational entropy reaching a maximum at the equiatomic composition in a binary system.

configurational entropy ($\Delta S_{\text{conf,max}} = R\ln N$) in any system increases with increasing number of elements (N) in the system (Figure 2.3). He empha-sized that the high mixing entropy (in this book "configurational entropy" is referred to as "mixing entropy/entropy of mixing" in some places to be in tune with literature. However, both these terms should be treated as same) in HEAs would have a profound effect on the constitu-ent phases, kinetics of phase formation, lattice strain, and thus properties. In particular, it enhances the mutual solubility between constituent com-ponents and leads to simpler phases and microstructure not expected before. This simplification originating from high-entropy effect is thus very important in such multicomponent alloys. Therefore, many possible new materials, new phenomena, new theories, and new applications are foreseen in the twenty-first century (Yeh et al., 2004b).

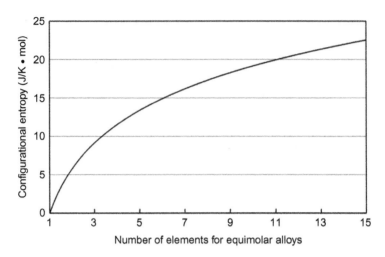

Figure 2.3 Entropy of mixing as a function of the number of elements for equiatomic alloys in the random solution state.

This concept has opened up flood gates of research activity and over the past decade about 400 papers have been published on HEAs, out of which about 100 are from Yeh's group itself. Though Yeh et al. (2004b) originally proposed that due to the large mixing entropy, HEAs tend to form solid solutions including disordered and partially ordered solutions intense research in recent years has been done on the rules or criteria using parameters including configurational entropy, enthalpy of mixing and the atomic size difference between different elements to predict the nature of phase formed in the alloy. Furthermore, computational thermodynamics methods such as CALPHAD have been used for predicting phase diagrams for HEAs in a more direct manner.

In addition to studying the basic phases and microstructure of different equiatomic multicomponent alloys ranging from five to twenty components (Cantor et al., 2004), Cantor et al. (2002) introduced a second strategy of equiatomic substitution of elements in binary alloys, wherein various similar elements are substituted in equal proportions in a binary system. This research was motivated by a desire to determine favorable compositions for the formation of metallic glasses (Figure 2.4). For example, in $Zr_{50}Cu_{50}$ alloy, he substituted 75% of Zr with Ti, Hf and Nb such that all four elements are in equiatomic proportions. Likewise, Cu was substituted partially with Ag and Ni so that all the three elements are in equiatomic proportions to arrive at a final septenary alloy composition of $(Ti_{0.25}Zr_{0.25}Hf_{0.25}Nb_{0.25})_{50}(Cu_{0.33}Ag_{0.33}Ni_{0.33})_{50}$. Even though this alloy has seven components, it can still be treated as a pseudobinary system and

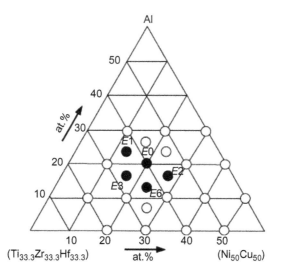

Figure 2.4 Ternary phase diagram for melt spun $(Ti_{33.3}Zr_{33.3}Hf_{33.3})_{100-x-y}(Ni_{50}Cu_{50})_xAl_y$ alloys, showing the influence of equiatomic substitution in a ternary system on glass forming ability (Kim et al., 2004).

Cantor and his group have demonstrated good glass forming ability in a number of such pseudobinary alloys. They have also added Al to the alloy to form pseudoternary alloys. Of course such substitution of like elements in place of the two components need not be in equal proportions and several metallic glasses are known with such nature, for example, an $Fe_{80}B_{20}$ glassy alloy, in which Fe can be substituted by a number of transition elements such as Cr, Co, and Ni and B can be substituted by other metalloids such as Si and C. One can demonstrate similar substitution even in case of intermetallic compounds (NiAl, wherein Ni can be substituted by Fe, Co) and even in oxides such as $BaTiO_3$, wherein Ba can be substituted by Sr and Ti can be substituted by Zr. Under such a substitution strategy with chemically similar elements, it is indeed very probable to generate similar or improved behavior and properties by using higher number of components.

Gibbs phase rule, which gives the degrees of freedom (F) a system has while maintaining equilibrium between a number of phases (P) containing a certain number of components (C), at constant pressure, can be expressed as

$$P + F = C + 1$$

Gibbs phase rule thus gives the maximum number of phases possible in a system of C components under equilibrium. The phase rule indicates that condensed systems (the system that does not have the gaseous phase) cannot have more than $C+1$ number of phases in equilibrium. Thus, binary, ternary, quaternary, and quinary systems cannot have more than 3, 4, 5, and 6 phases in equilibrium, respectively. However, such a condition of finding the maximum number of phases is observed in any system under a specific condition wherein the degrees of freedom are zero (invariant condition). This does not mean that the system cannot have less number of phases and for example, a quinary condensed system can have any number of phases in equilibrium ranging from 1 to 6 at various compositions and temperatures.

However, it is important to note that the number of phases observed in these HEAs is significantly less than the maximum number of phases expected from the phase rule, suggesting that large configurational entropy in these alloys enhances the mutual solubility for forming solid solution phases (disordered or partially ordered) and thus restricts the formation of a large number of phases. Furthermore, HEAs being multiprincipal-element alloys, the diffusivities of atoms are expected to be low and hence the formation of a number of phases is kinetically constrained in such alloys. Thus, the observation of less number of phases in HEAs is not only due to high configurational entropy of these alloys but also due to low diffusivities in these multicomponent alloys. These merits are often useful in controlling microstructure and properties of a material for practical applications.

2.3 DEFINITION OF HEAs

Each high-entropy alloy contains multiple elements, often five or more in equiatomic or near-equiatomic ratios, and minor elements (Yeh et al., 2004b). The basic principle behind HEAs is that significantly high mixing entropies of solid solution phases enhance their stability as compared with intermetallic compounds, especially at high temperatures. This enhancement allows them to be easily synthesized, processed, analyzed, manipulated, and utilized by us. In a broad sense, HEAs are preferentially defined as those alloys containing at least five principal elements, each having the atomic percentage between 5% and 35%. The atomic percentage of each minor element, if any, is hence less than 5% (Yeh et al., 2004b).

Why are such multiprincipal-element alloys called HEAs? From statistical thermodynamics, Boltzmann's equation (Gaskell, 1995; Swalin, 1972) can be used for calculating configurational entropy of a system:

$$\Delta S_{conf} = k \ln w$$

where k is Boltzmann's constant and w is the number of ways in which the available energy can be mixed or shared among the particles of the system. Thus the configurational entropy change per mole for the formation of a solid solution from n elements with x_i mole fraction is:

$$\Delta S_{conf} = -R \sum_{i=1}^{n} X_i \ln X_i$$

Let us consider an equiatomic alloy at its liquid state or regular solid solution state. Its configurational entropy per mole can be calculated as follows (Yeh et al., 2004b; Yeh, 2006):

$$\Delta S_{conf} = -k \ln w = -R\left(\frac{1}{n}\ln\frac{1}{n} + \frac{1}{n}\ln\frac{1}{n} + \cdots \frac{1}{n}\ln\frac{1}{n}\right)$$

$$= -R \ln\frac{1}{n} = R \ln n$$

where R is the gas constant, 8.314 J/K mol.

Figure 2.5 shows an example illustrating the formation of quinary equiatomic alloy from five elements. From the above equation, ΔS_{conf} can be calculated as $R \ln 5 = 1.61R$. For a nonequiatomic HEA, the mixing entropy would be lower than that for an equiatomic alloy. Consider a nonequiatomic alloy $Al_{1.5}CoCr_{0.5}FeNi_{0.5}$ (or $Al_{33.3}Co_{22.2}Cr_{11.1}Fe_{22.2}Ni_{11.1}$ in at.%). Its configurational entropy can be calculated as $1.523R$ which is slightly smaller than $1.61R$ of the equiatomic alloy AlCoCrFeNi.

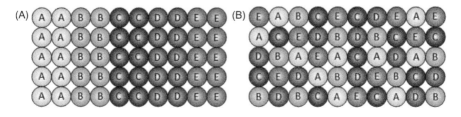

Figure 2.5 (A) Five components in equiatomic ratio before mixing and (B) mixing to form a random solid solution. Assume their atomic sizes are same.

Table 2.1 Configurational Entropies of Equiatomic Alloys with Constituent Elements up to 13												
N	1	2	3	4	5	6	7	8	9	10	11	12	13
ΔS_{conf}	0	0.69R	1.1R	1.39R	1.61R	1.79R	1.95R	2.08R	2.2R	2.3R	2.4R	2.49R	2.57R

Although total mixing entropy has four contributions such as configurational, vibrational, magnetic dipole, and electronic randomness, configurational entropy is dominant over other three contributions (Fultz, 2010; Swalin, 1972). Among the other three, the excess vibrational entropy due to mixing at high temperatures could be calculated from Debye temperatures (Swalin, 1972). Negative contribution of excess vibrational entropy to overall mixing entropy might occur and could be enhanced by attractive interactions between unlike atomic pairs. Table 2.1 lists the configurational entropies of equiatomic alloys with increasing number of components in terms of gas constant R. The entropy increases as the number of element increases. From Richards' rule, $\Delta H_f / T_m = \Delta S_f \sim R$, the entropy change per mole, ΔS_f, from solid to liquid during melting is about one gas constant R, and the enthalpy change or latent heat per mole, ΔH_f, can be estimated as RT_m, where T_m is the melting point. Because ΔH_f can be regarded as the energy required to destroy about one twelfth of all bonds in the solid, the mixing entropy of R per mole in random solid solution is quite large to lower its mixing free energy, ΔG_{mix}, since $\Delta G_{mix} = \Delta H_{mix} - T\Delta S_{mix}$ by the amount of RT. For example, the amount at 1000 K is $RT = 8.314$ kJ/mol. It is also well known that an ideal monoatomic gas (including metals) per mole has a kinetic energy or internal energy of $1.5RT$. Therefore, such a free energy lowering causes the solid solution phases to have a greater ability to compete with intermetallic compounds, which usually have much lower ΔS_{conf} due to their ordered nature. That means the tendency to form the mixing state of constituent elements would be increased by increased mixing entropy, especially at high temperatures.

It can be seen from Table 2.1 and Figure 2.3 that the configurational entropy of ternary equiatomic alloy is already slightly higher than $1R$ and that of quinary alloy is higher than $1R$ by 61%. Furthermore, the formation enthalpies of two typical strong intermetallic compounds, NiAl and TiAl, divided by their respective melting points, lead to $1.38R$ and $2.06R$, respectively (Yeh et al., 2004b). Thus, it

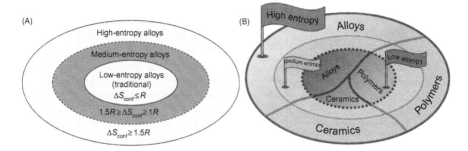

Figure 2.6 (A) Alloys world (Yeh, 2013a) and (B) materials world (excluding composite materials) based on configurational entropy.

is reasonable to think that a ΔS_{conf} of $1.5R$ (even not including other three mixing entropy contributions mentioned above) is large enough against the strong ordering of such atomic pairs with strong bonding energies at high temperatures. The ΔS_{conf} of $1.5R$ is recommended as a borderline between high- and medium-entropy alloys. In addition, $1R$ is recommended as the borderline for medium- and low-entropy alloys because a mixing entropy smaller than $1R$ is expected to have less power to compete with those strong bonding energies (Yeh, 2013a). Based on this, the alloys world is schematically shown in Figure 2.6. Similar mixing entropy effect may be expected in ceramics and polymers. Larger number of components in ceramics and polymers is expected to yield higher mixing entropy effect. Therefore, as shown in Figure 2.6B, ceramics can be grouped into three categories as low-, medium-, and high-entropy ceramics (HECs). Similarly, polymers also have three similar categories. Previously, borderlines for alloys world were set as 0.69 and $1.61R$, respectively, and thus have some difference from Figure 2.6A (Yeh, 2006, 2013a; Yeh et al., 2007b). This is because the previous ones are based on the number of principal elements, i.e. 2 and 5, respectively.

Table 2.2 gives the configurational entropies calculated for typical traditional alloys in their liquid state or random state. From this table, most alloys have low entropy (smaller than $1R$) whereas some concentrated alloys of Ni-base, Co-base superalloys, and BMGs (bulk metallic glasses) have medium entropy between 1 and $1.5R$. That means none of the traditional alloys have high mixing entropy $\geq 1.5R$.

Based on the above, the composition definition of HEAs in a broad sense can be explained to some extent from configurational entropy

Table 2.2 Configurational Entropies Calculated for Typical Traditional Alloys at Their Liquid State or Random State (Yeh, 2013a)

Systems	Alloys	ΔS_{conf} at Liquid State
Low-alloy steel	4340	$0.22R$ low
Stainless steel	304	$0.96R$ low
	316	$1.15R$ medium
High-speed steel	M2	$0.73R$ low
Mg alloy	AZ91D	$0.35R$ low
Al alloy	2024	$0.29R$ low
	7075	$0.43R$ low
Cu alloy	7−3 brass	$0.61R$ low
Ni-base superalloy	Inconel 718	$1.31R$ medium
	Hastelloy X	$1.37R$ medium
Co-base superalloy	Stellite 6	$1.13R$ medium
BMG	$Cu_{47}Zr_{11}Ti_{34}Ni_8$	$1.17R$ medium
	$Zr_{53}Ti_5Cu_{16}Ni_{10}Al_{16}$	$1.30R$ medium

calculation. From the configurational entropy equation, we can observe that an element having a concentration of 5 at.% would contribute a mixing entropy of $0.05R \ln 0.05 = 0.15R$, which is just 10% of minimum requirement of $1.5R$ for HEAs. So, we could regard an element in an amount ≥ 5 at.% as a principal element. For elements with concentrations of 4, 3, 2, and 1 at.%, the contributions are 0.129R, 0.105R, 0.078R and 0.046R, respectively. So, we regard an element with $< 5\%$ as minor element (Yeh, 2013a).

There arises a question from the definition: what is the upper bound of the number of metallic principal elements? For 5-, 10-, 12-, 13-, 14-, 15-, 20-, and 40-component equiatomic alloys, the total configurational entropies are $1.61R$, $2.3R$, $2.49R$, $2.57R$, $2.64R$, $2.71R$, $3.0R$, and $3.69R$, respectively. Beyond the 13 component alloy system, the increase is only 0.07R for each additional element, which is relatively small. Hence a practical number of principal element between 5 & 13 was suggested (Yeh et al., 2004b; Yeh, 2013a). That means more principal elements will not get significant benefit from the high-entropy effect but might increase the complexity in handling raw materials or recycling the alloys.

Indeed, it is not easy to give a clear-cut composition definition for HEAs. The composition definition is just a guideline. An alloy with some deviation from this composition definition might still be regarded as a HEA. For example, an alloy with 21-component equiatomic alloy is surely a HEA, even though each element has a concentration smaller than 5 at.%. Miracle et al. (2014) suggest that the definition, $S_{conf} \geq 1.5R$, based on configurational entropy change is an operational definition of HEAs since the entropy change can be calculated for any alloy with an ideal or regular configurational entropy at the random solid solution state. In addition, this definition has more information and is more consistent than definitions based on alloy composition or on the value of S_{conf} at low temperatures. This operational definition also includes HEAs with two or more phases at low temperatures.

If we use the above composition definition and the practical upper bound of 13 principal elements, at an arbitrary choice of a group of 13 metallic elements, we can obtain a total of 7099 possibilities for designing equiatomic HEAs from 5 to 13 elements (Yeh et al., 2004b):

$$C_5^{13} + C_6^{13} + C_7^{13} + C_8^{13} + C_9^{13} + C_{10}^{13} + C_{11}^{13} + C_{12}^{13} + C_{13}^{13} = 7099$$

For any system, we may design equiatomic HEAs like AlCoCrCuFeNi. We may also design nonequiatomic HEAs with minor alloying elements like $AlCo_{0.5}CrCuFe_{1.5}Ni_{1.2}B_{0.1}C_{0.15}$ for further modification of microstructure and properties. Since the number of common metals is much larger than 13, HEA systems are numerous and HEAs are even more in number.

2.4 COMPOSITION NOTATION

As the composition of each HEA does not contain a single principal element, the sequence of the composing elements in the chemical formula could be shown in many different ways. For the sake of easy identification and to make comparisons between alloys of the same system or alloys belonging to different systems, one common method is to put the element name of principal elements in alphabetical sequence, and finally put minor elements behind but in similar alphabetical sequence. This is convenient but does not carry additional meaning. Ordering the elements by atomic number or Mendeleev number may have more significance but lacks the convenience of alphabetical order.

The concentration could be expressed as atomic ratio, or atomic percentage in the subscript position. For example, the expressions for two equiatomic alloys, AlCoCrFeNiTi and CoCrCuFeMnMoNiZr, conform to this rule.

The expressions for two nonequiatomic alloys, $Al_{0.5}Co_{1.5}CrFeNi_{1.5}Ti_{0.5}$ and $CoCr_{0.5}Cu_{0.5}Fe_{1.5}Mn_{0.5}Mo_{0.3}NiZr_{0.4}$, in atomic ratio also follow this rule. These two compositions could also be expressed as $Al_{8.3}Co_{25}Cr_{16.7}Fe_{16.7}Ni_{25}Ti_{8.3}$ and $Co_{17.5}Cr_{8.8}Cu_{8.8}Fe_{26.3}Mn_{8.8}Mo_{5.3}Ni_{17.5}Zr_{7.0}$ in at.%. In addition, the two nonequiatomic alloys with minor additions can be expressed as $Al_{0.5}Co_{1.5}CrFeNi_{1.5}Ti_{0.5}B_{0.1}C_{0.2}$ and $CoCr_{0.5}Cu_{0.5}Fe_{1.5}Mn_{0.5}Mo_{0.2}NiZr_{0.2}C_{0.1}Si_{0.15}$ in atomic ratio. Molar ratio has also been used in HEA literature simply because molar ratio is equal to atomic ratio (the number of atoms in one mole of HEAs is an Avogadro's number). However, this rule need not be strictly followed if some composition feature is to be emphasized. A suitable expression can be used to reveal the specialty. For example, the notation $Al_{0.5}Co_xCrFeNi_{1.5}Ti_{0.5}B_{0.1}C_{0.2}$ ($x = 0.5, 0.75, 1.0, 1.25, 1.5$) represents five compositions in which Co is varied from 0.5 to 1.5.

HECs have also been developed and investigated based on the similar alloy concept of HEAs. Their components corresponding to elements in HEAs are binary nitrides (e.g., AlN, CrN, Si_3N_4, TiN), binary carbides (e.g., Cr_3C_2, SiC, TiC, VC), binary borides (e.g., CrB, NbB_2, TaB, TiB_2). The compositions of HECs are also preferentially defined as those ceramics containing at least five principal components, each having a mole percentage between 5% and 35%. The free energy of mixing of a HEC is the change between the free energy of the HEC and the overall free energy of components in their standard state before mixing. The enthalpy of mixing and entropy of mixing are also similar changes. The binary compound has no mixing entropy and enthalpy. The mixing entropy can be calculated by the mixing of elements since N atoms occupy one set of sublattice whereas metal atoms randomly occupy another set of sublattice. The mixing enthalpy is mainly from lattice distortion energy or lattice strain energy since the chemical bonding between metals and nitrogen after mixing similarly can be found in the unmixed state. Their composition formula in stoichiometric ratio could be (Al,Cr,Ta,Ti,Zr)N in atomic ratio, or (Al,Cr,Ta,Ti,Zr)$_{50}$N$_{50}$ in atomic percentage. If the metal elements are known to be in equiatomic ratio, then (AlCrTaTiZr)N in atomic ratio, or

$(AlCrTaTiZr)_{50}N_{50}$ in atomic percentage could be used. On the other hand, the formula in nonstoichiometric ratio could be $(Al,Cr,Ta,Ti,Zr)_{1-x}N_x$ in atomic ratio, or $(Al,Cr,Ta,Ti,Zr)_{100-y}N_y$ in atomic percentage. If carbon is involved, it could have the form $(Al,Cr,Ta,Ti,Zr)_{1-u-v}N_uC_v$ or $(Al,Cr,Ta,Ti,Zr)_{100-s-t}N_sC_t$.

2.5 FOUR CORE EFFECTS OF HEAs

There are many factors affecting microstructure and properties of HEAs. Among these, four core effects are most basic (Yeh, 2013a). Because HEAs contain at least five major elements, and conventional alloys are based on one or two metal elements, different basic effects exist between these two categories. The four core effects are high-entropy, severe lattice distortion, sluggish diffusion, and cocktail effects (Yeh, 2006, 2013a). For thermodynamics, high-entropy effect could interfere with complex phase formation. For kinetics, sluggish diffusion effect could slow down phase transformation. For structure, severe lattice distortion effect could alter properties to an extent. For properties, a cocktail effect brings excess to the quantities predicted by the mixture rule due to mutual interactions of unlike atoms and severe lattice distortion.

2.5.1 High-Entropy Effect

High-entropy effect is the most important effect because it can enhance the formation of solid solutions and makes the microstructure much simpler than expected. Based on physical metallurgy concepts, we easily expect such alloys to have many different interactions among elements and thus form many different kinds of binary, ternary, quaternary compounds, and/or segregated phases. Thus, such alloys would possess complicated structure not only difficult to analyze but also brittle by nature. This expectation in fact neglects the effect of high mixing entropy.

As for the mixing enthalpy, ΔH_{mix}, it can be calculated as (de Boer et al., 1988; Takeuchi and Inoue, 2000, 2005):

$$\Delta H_{mix} = 4 \sum_{i=1, j \neq i}^{n} \Delta H_{\langle ij \rangle}^{mix} X_i X_j + \sum_{k} \Delta H_k^{trans} X_k$$

where x_k is the mole fraction of nonmetallic element k in the system and $\Delta H^{\text{mix}}_{(ij)}$ is the mixing enthalpy per mole of an equiatomic $i - j$ alloy in the solid state. The transformation enthalpies of k-element from its standard state to metallic state are 30, 180, 310, 37, 25, and 17 kJ/mol for B, C, N, Si, Ge, and P, respectively.

As the Gibbs free energy of mixing, ΔG_{mix}, is

$$\Delta G_{\text{mix}} = \Delta H_{\text{mix}} - T \Delta S_{\text{mix}}$$

where ΔH_{mix} and ΔS_{mix} are the enthalpy of mixing and entropy of mixing, respectively, higher number of element would potentially lower the mixing free energy, especially at high temperatures by contributing larger ΔS_{mix}.

In the solid state of an alloy, although there are numerous possible states, the equilibrium state is the one having the lowest free energy of mixing according the second law of thermodynamics. There are three possible categories of competing states, that is, elemental phases, intermetallic compounds and solid solution phases below the lowest melting point of the alloy (Yeh, 2013a). Elemental phase is the terminal solid solution based on one metal element. Intermetallic compound is a stoichiometric compound having specific superlattices, such as NiAl having B2 structure and Ni_3Ti having $D0_{24}$ structure. Solid solution is the phase with the complete mixing or significant mixing of all elements in the structure of BCC (body-centered cubic), FCC, or HCP (hexagonal close-packed). Intermetallic phases or intermediate phases based on intermetallic compounds are also regarded as solid solutions, but they are partially ordered solid solutions (Cullity and Stock, 2001; Reed-Hill and Abbaschian, 1994). In such phases, different constituent elements tend to occupy different set of lattice sites.

In order to reveal the high-entropy effect which enhances the formation of solid solution phases and inhibits the formation of intermetallic compounds, consider a high-entropy alloy composed of those constituent elements with stronger bonding between each other. If strain energy contribution (due to atomic size difference) to mixing enthalpy is neglected for simplicity, the free energies of mixing for various kinds of state are shown in Table 2.3 (Yeh, 2013a). Elemental phases have small negative ΔH_{mix} and ΔS_{mix} because they are based on one major element. Compound phases have large negative ΔH_{mix} but small ΔS_{mix} because ordered structures have small configurational entropy. But random solid

Table 2.3 Comparisons of ΔH_{mix}, ΔS_{mix}, and ΔG_{mix} Among Elemental Phases, Compounds, Ordered Solid Solutions, and Random Solid Solutions for n-element HEAs with Stronger Bonding Between Composing Elements

Comparative States	Elemental Phases	Compounds	Intermediate Phases	Random Solid Solutions
ΔH_{mix}	~ 0	Large negative	Less large negative	Medium negative
ΔS_{mix}	~ 0	~ 0	Medium	$\Delta S_{mix} = -R\sum_{i=1}^{n} X_i \ln X_i$
ΔG_{mix}	~ 0	Large negative	Larger negative	Larger negative

Strain energy from atomic size difference is not included in ΔH_{mix} (Yeh, 2013a).

solution phases containing multicomponents have medium negative ΔH_{mix} and highest ΔS_{mix}. Why multicomponent solid solutions have medium ΔH_{mix}? This is because there exists a proportion of unlike atomic pairs in solution phases. For example, a mole of atoms, N_0, of NiAl intermetallic compound (B2) in complete ordering has $(1/2) \times 8N_0$ Ni—Al bonds (coordination number in ordered BCC is 8), whereas a mole of NiAl random solid solution would have $(1/2) \times (1/2) \times 8N_0$ Ni—Al bonds.

That means the mixing enthalpy in the random solution state is half that of the completely ordered state. Assuming that all heats of mixing for unlike atom pairs are the same, ΔH_{mix} of the random solution state for quinary equiatomic alloy is 4/5 of that of its completely ordered state. Similarly, for septenary equiatomic alloy, the ratio is 6/7 (~ 0.86). Therefore, a higher number of elements would allow the random solution state to have the mixing enthalpy much closer to that of the completely ordered state. With the aid of its high mixing entropy in lowering the overall mixing free energy, random solution state would be more favorable in thermal stability than the ordered state. Although not all heats of mixing for unlike atom pairs are the same in reality, the tendency to increase the degree of disorder at equilibrium is unchanged. That is, at least, partially ordered solid solution phases having multicomponent composition and a certain degree of disorder instead of stoichiometric compounds are favorable at equilibrium. This is the reason why the state of intermediate phases as shown in Table 2.3 could be often favorable. Obviously this tendency toward disordered state is stronger at higher temperature due to $-T\Delta S_{mix}$ effect. Miracle et al. (2014) have made a rough,

Table 2.4 Binary Mixing Enthalpies, ΔH^{mix}_{ij} (kJ/mol), for Unlike Atom Pairs in Co–Cr–Fe–Mn–Ni Alloys (de Boer et al., 1988)				
Cr	−3.8	5.9	6.8	6.4
−3.8	Mn	−4.7	2.2	−14
5.9	−4.7	Fe	−1.3	−4.6
6.8	2.2	−1.3	Co	0.0
6.4	−14	−4.6	0.0	Ni

order-of-magnitude thermodynamic analysis to demonstrate this effect. The analysis suggests that ΔS_{conf} of HEAs may be sufficient to destabilize 5−10% of intermetallic compounds (those with the lowest enthalpies of formation) at room temperature. An additional 30−55% of ordered compounds may be suppressed in HEAs at 1500 K. Roughly 50% of the intermetallic compounds may be stable at 300 K but unstable at 1500 K. They also pointed out that this effect also offers a new approach for controlling microstructure (via particle dissolution and subsequent controlled precipitation) and properties in particulate-strengthened HEAs.

Consider another case where one HEA contains an element which has strong repulsion force to other elements, that is, it has large positive mixing enthalpy with other elements. It is expected that the element will segregate to form an elemental phase with little solubility with other elements. Mixing entropy effect is still not sufficient to overcome the large enthalpy effect. Suppose the element has weak repulsion force, that is, has small positive mixing enthalpy, with some other elements, its segregation phase will contain higher concentration of these elements due to mixing entropy effect and becomes a concentrated solid solution with a major element.

In general, if the heats of mixing for unlike atomic pairs have smaller difference, simple solid solution phases would be dominant in the equilibrium state. For example, CoCrFeMnNi alloy has small difference in such heats of mixing as shown in Table 2.4 (de Boer et al., 1988) and thus form a simple FCC solution even full annealed at all temperatures (Otto et al., 2013a; Tsai et al., 2013b). Refractory HfNbTaTiZr alloy has small difference as shown in Table 2.5 (de Boer et al., 1988) and can form a simple BCC phase in the as-cast state (Senkov et al., 2011a). Conversely, large difference among the heats of

Table 2.5 Binary Mixing Enthalpies, ΔH^{mix}_{ij} (kJ/mol), for Unlike Atom Pairs in Hf–Nb–Ta–Ti–Zr Alloys (de Boer et al., 1988)

Hf	4	3	0	0
4	Nb	0	2	4
3	0	Ta	1	3
0	2	1	Ti	0
0	4	3	0	Zr

Table 2.6 Binary Mixing Enthalpies, ΔH^{mix}_{ij} (kJ/mol), for Unlike Atom Pairs in Al–Co–Cr–Cu–Fe–Ni Alloys (de Boer et al., 1988)

Al	−19	−10	−1	−11	−22
−19	Co	−4	6	−1	0
−10	−4	Cr	12	−1	−7
−1	6	12	Cu	13	4
−11	−1	−1	13	Fe	2
−22	0	−7	4	2	Ni

mixing for unlike atomic pairs might generate more than two phases. For example, Al has stronger bonding with Ni, Co, Fe, and Cr but Cu has repulsive bond with Fe, Cr, and Co as shown in Table 2.6 (de Boer et al., 1988). As a result, equiatomic AlFeCoCrCuNi alloy forms Cu-rich FCC + multicomponent FCC + multicomponent BCC (A2) at high temperatures above 600°C, and has B2 precipitates in the Cu-rich FCC and spinodally decomposed structure of A2 + B2 phases from A2 phase during cooling. B2 solid solution containing multicomponents is in fact the intermediate phase derived from NiAl-type compound (Zhang et al., 2008b).

The atomic size difference among elements in a HEA also affects the stability of disordered solid solution phases. This is because large atomic size difference would cause the lattice to be heavily distorted and contribute strain energy to mixing enthalpy and thus mixing free energy. On the other hand, there are a variety of structures which preferentially form under large atomic size difference. It has been known that atomic size is one of the main factors which determine the structure of intermetallic compounds or intermediate phases (Porter and Easterling, 1992; Massalski, 1989, Pettifor, 1996; de Graef and

McHenry, 2012). For example, when the component atoms differ in size by a factor of about 1.1−1.6, it is possible for these atoms to fill space most efficiently by the arrangement into ordered structures based on $MgCu_2$, $MgZn_2$, and $MgNi_2$, that is, the so-called Laves phases. Another case is that when very small atoms such as H, B, C, and N are incorporated, interstitial compounds might form in the formulas MX, M_2X, MX_2, and M_6X, where M can be Zr, Ti, V, Cr, etc. and X represents the small atoms. Therefore, it can be said that large atomic size difference might enhance the formation of different compounds or intermediate phases.

2.5.2 Severe Lattice Distortion Effect

Because the multicomponent matrix of each solid solution phase in HEAs is a whole-solute matrix, every atom is surrounded by different kinds of atom and thus suffers lattice strain and stress mainly due to the atomic size difference as shown in Figure 2.7, an example with 10 kinds of atoms. The average lattice exists because it can be determined by x-ray diffraction. Besides atomic size difference, different bonding energy and crystal structure among constituent elements are also expected to cause even higher lattice distortion in consideration of the nonsymmetrical neighboring atoms, that is, nonsymmetrical

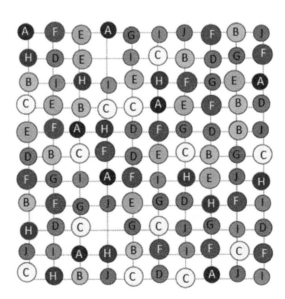

Figure 2.7 Two-dimensional matrix of a solid solution with 10 different components. Two vacancies are presented. Average lattice is shown by dotted lines.

One-component alloy   Five-component alloy

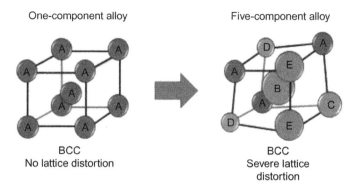

BCC   BCC
No lattice distortion   Severe lattice
                                                distortion

Figure 2.8 Schematic diagram showing large lattice distortion exists in the five-component BCC lattice.

bindings and electronic structure, around an atom, and the variation of such nonsymmetry from site to site (Yeh et al., 2007a; Tsai et al., 2013b). Thus, most lattice sites and overall lattice distortion would be more severe than that in conventional alloys in which most matrix atoms (or solvent atoms) have the same kind of atoms as their neighbors. Figure 2.8 shows the severe lattice distortion of a five-component BCC lattice. The three-dimensional unit cell is highly distorted.

Lattice distortion not only affects various properties but also reduces the thermal effect on properties. Hardness and strength effectively increase because of large solution hardening in the heavily distorted lattice. For example, Vickers hardness of FCC equiatomic alloy CoCrFeMnNi is 1192 MPa in the homogenized state which is higher than that 864 MPa obtained by the mixture rule. Hardness value of refractory BCC equiatomic alloy MoNbTaVW is 5250 MPa which is three times that obtained by the mixture rule (Senkov et al., 2010). It appears that FCC alloys display much smaller solution hardening than BCC alloys. This might be due to that FCC lattice has 12 nearest neighbors whereas BCC one has 8 nearest neighbors. For the same set of elements, FCC lattice has smaller fraction of unlike atomic pairs in the first nearest neighbors and thus a smaller distortion strain and solution hardening than BCC lattice. Besides, CoCrFeMnNi alloy has smaller atomic size difference than MoNbTaVW and thus smaller total distortion.

For the x-ray diffraction peak intensity, the distorted atomic planes increase x-ray diffuse scattering effect and give smaller peak intensity (Yeh et al., 2007a). Lattice distortion also causes electron scattering and significant decrease in electrical conductivity. This in turn reduces the electron contribution to thermal conductivity by electron conduction. Phonon scattering also becomes larger in the distorted lattice and decrease the thermal conductivity (Kao et al., 2011). All these properties in HEAs are found to be quite insensitive to temperature. For example, the temperature coefficients of resistivity are quite small in HEAs. This is because of the fact that the lattice distortion caused by thermal vibration of atoms is relatively small as compared with the severe lattice distortion (Yeh et al., 2007a).

2.5.3 Sluggish Diffusion Effect

In diffusion-controlled phase transformation, the formation of new phases requires cooperative diffusion of many different kinds of atoms to accomplish the partitioning of composition in HEAs. As explained in the last section, a HEA mainly contains random solid solution, and/or ordered solid solution. Their matrices could be regarded as whole-solute matrices. As a result, the diffusion of an atom in a whole-solute matrix would be very different from that in the matrix of conventional alloys. A vacancy in the whole-solute matrix is in fact surrounded and competed by different-element atoms during diffusion. It has been proposed that slower diffusion and higher activation energy would occur in HEAs due to larger fluctuation of lattice potential energy (LPE) between lattice sites (Tsai et al., 2013b). The abundant low-LPE sites can serve as traps and hinder the diffusion of atoms. This leads to the sluggish diffusion effect.

Tsai et al. (2013b) made diffusion couples from a near-ideal solution system of Co–Cr–Fe–Mn–Ni to analyze the diffusion data of each element in the matrix. The results concluded that the sequence in the order of decreasing diffusion rate was Mn, Cr, Fe, Co, and Ni. Diffusion coefficients of each elements at T/T_m in the Co–Cr–Fe–Mn–Ni alloy system were the smallest in comparison to similar FCC matrices including Fe–Cr–Ni(–Si) alloys and pure Fe, Co, and Ni metal. In addition, the melting point normalized activation energies, Q/T_m, in the HEA were the largest as shown in Figure 2.9. Also note that for the same element, the degree of sluggish diffusion is related to the number of elements in that matrix. For example, the Q/T_m values in the present HEAs are the highest; those in Fe–Cr–Ni(–Si) alloys are the second; and those in pure metals are the lowest. This suggests that greater the number of elements, slower is the diffusion rate (Tsai et al., 2013b).

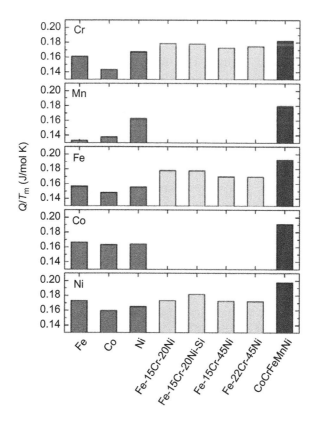

Figure 2.9 Comparison among the melting point normalized activation energy of diffusion for Cr, Mn, Fe, Co, and Ni in different matrices: pure metals, stainless steels, and high-entropy alloy CoCrFeMnNi (Tsai et al., 2013b).

It is expected that sluggish diffusion might affect phase nucleation, growth and distribution, and morphology of new phase through diffusion-controlled phase transformation. It also provides various advantages in controlling microstructure and properties: easiness to get supersaturated state and fine precipitates, increased recrystallization temperature, slower grain growth, reduced particle coarsening rate, and increased creep resistance. As an example, Liu et al. (2013) found a significantly high value of activation energy ($\sim$ 321 kJ/mol) and slow coarsening kinetics on the grain growth behavior of CoCrFeMnNi HEA following cold rolling and annealing, which is in agreement with the sluggish diffusion effect. As a result, sluggish diffusion is in general positive for improving properties of HEAs. For example, fine precipitation and grain structure could improve the combination of strength and toughness. Improved creep resistance could prolong the life of parts used at high temperatures.

2.5.4 Cocktail Effect

The article on "Alloyed pleasures: multimetallic cocktails" by Ranganathan (2003) caught the imagination of scientists in this area and the paper attracted about 100 citations in the past decade. In HEAs, cocktail effect is used to emphasize the enhancement of the properties by at least five major elements. Because HEAs might have single phase, two phases, three phases, or more depending on the composition and processing, the whole properties are from the overall contribution of the constituent phases. This relates with the phase size, shape, distribution, phase boundaries, and properties of each phase. Moreover, each phase is a multicomponent solid solution and can be regarded as an atomic-scale composite. Its composite properties not only come from the basic properties of elements by the mixture rule but also from the mutual interactions among all the elements and from the severe lattice distortion. Mutual interaction and lattice distortion would bring excess quantities in addition to those predicted by the mixture rule. As a whole, "cocktail effect" ranges from atomic-scale multicomponent composite effect to microscale multiphase composite effect. Figure 2.10 (Yeh, 2006) shows the cocktail effect introduced by the interaction of constituent elements in the $Al_x CoCrCuFeAl$ alloy, which leads to a phase transformation from an FCC structure to a BCC when the Al content is increased beyond critical value.

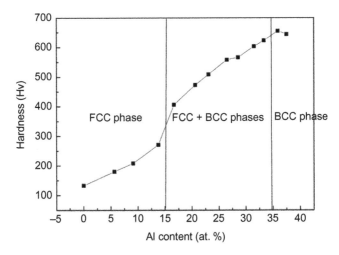

Figure 2.10 Cocktail effect introduced by the interaction of constituent elements in the $Al_x CoCrCuFeAl$ alloy.

To illustrate the cocktail effect, two examples are introduced. In the first example, refractory HEAs developed by Air Force Research Laboratory have melting points very much higher than that of Ni-base and Co-base superalloys (Senkov et al., 2010, 2011b). They mainly consist of refractory elements having melting temperatures above 1650°C. By the mixture rule, quaternary equiatomic alloy MoNbTaW and quinary equiatomic alloy MoNbTaVW have melting points above 2600°C. They actually found that both alloys display much higher softening resistance than superalloys and have yield strength above 400 MPa at 1600°C (Senkov et al., 2011a). Such refractory HEAs are thus expected to have promising potential in those applications at very high temperatures. In the second example, Zhang et al. (2013b) studied $CoNiFe(AlSi)_{0-0.8}$ alloys for obtaining the optimum combination of magnetic, electrical, and mechanical properties. The best alloy is $CoNiFe(AlSi)_{0.2}$ with saturation magnetization (1.15 T), coercivity (1400 A/m), electrical resistivity (69.5 $\mu\Omega$-cm), yield strength (342 MPa), and strain without fracture (50%). This demonstrates that the alloy is an excellent soft magnetic material for many potential applications (Zhang et al., 2013b). This finding was based on suitable alloy design which used equiatomic ferromagnetic elements (Fe, Co, and Ni) for forming ductile FCC phase (atomic packing density higher than BCC), and suitable addition of nonmagnetic elements (Al and Si having slightly antiparallel magnetic coupling with Fe, Co, and Ni) to increase lattice distortion. As a result, it brought up a positive cocktail effect in achieving high magnetization, low coercivity, good plasticity, high strength, and high electrical resistance. Therefore, it is important for an alloy designer to understand related factors involved before selecting suitable composition and processes.

CHAPTER 3

Phase Selection in High-Entropy Alloys

3.1 PREDICTING SOLID SOLUBILITY FROM HUME-ROTHERY RULES

HEAs are essentially multicomponent equi-atomic or near equiatomic alloys which form mostly solid solutions including random solid solutions and partially ordered ones. In a few cases they form amorphous alloys. The results from the large number of papers published in this field indicate that the formation of a single-phase solid solution in these alloys is decided to a large extent by the proper choice of elements used for making the alloy.

This takes us back to Hume-Rothery (H-R) rules that were developed to identify conditions for any element to be soluble in another (Hume-Rothery, 1967). In the 1920s, Hume-Rothery identified the factors that influence compound formation and control alloying behavior, after surveying the available solubility data in detail. He identified that there is a connection between solubility and factors including atomic size, crystal structure, valence, and electronegativity of the two components of an alloy. In spite of a number of other researchers attempting to address this question of solubility, the simplicity and generality of H-R rules have become one of the important cornerstones in materials science (Massalski, 1989). The H-R rules for binary substitutional solid solutions are generally stated as follows (Smith and Hashemi, 2006):

1. The radii of the solute and solvent atoms must not differ by more than about 15%: For complete solubility, the atomic size difference should be less than 8%.
2. The crystal structures of the two elements must be the same for extended solid solubility.
3. Extended solubility occurs when the solvent and solute have the same valency.
4. The two elements should have similar electronegativity so that intermetallic compounds will not form.

These rules need revisiting in the context of multicomponent solid solutions, when interactions among the different components need to be taken into account.

H-R rules were given with a diagrammatic representation by Darken and Gurry (1953) (DG method), wherein they plotted the electronegativity and atomic size factor in a 2D map and identified an ellipse with atomic size factor of ±15% and electronegativity difference of ±0.4 to separate soluble elements from insoluble ones. They applied this to Mg, Ag, and Al systems and found that this approach worked well for Mg alloys systems but not for the Ag systems. Figure 3.1 shows such a Darken-Gurry map for Pd as a solvent (Fang et al., 2002). Gschneidner (1980) extended the DG method by introducing crystal structure (electronic crystal structure Darken−Gurry method, ECSDG) and was able to show improvement in the prediction of extensive solubility in different solvents (Mg, Al, Fe, Ge, Pd, Ag, Cd, La, W, and Pb) in comparison to DG method.

In 1970s Miedema and his coworkers (Miedema, 1973) were able to successfully calculate heats of formation of intermetallic alloys using two parameters namely, electronegativity difference ($\Delta\phi^*$) and the difference in electron density at the boundary of the Wigner−Seitz cell (Δn_{ws}) of pure metals. Chelikowsky (1979) introduced graphical representation of these two parameters to predict the solubility in case of divalent solvents

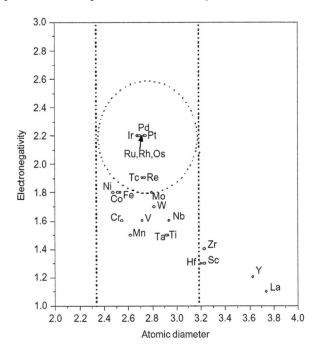

Figure 3.1 Darken−Gurry map for predicting the elements that are soluble in Pd (Fang et al., 2002).

and demonstrated better prediction capability. Alonso and Simozar (1980) extended this approach further by bringing in the atomic size also into the picture in addition to electronegativity and electron cell-boundary density and explained the degree of solubility of different elements in a given host element with a fair degree of success. This approach uses the combination of electronegativity and electron density into single parameter, ΔH_c (the heat of formation of an equiatomic compound), as proposed by Miedema, as given below:

$$\Delta H_c = -P(\Delta \phi^*)^2 + Q(\Delta n_{ws}^{1/3})^2 - R$$

where P, Q, and R are constants.

Incidentally, all these early efforts were restricted to binary and to some ternary systems.

3.2 MUTUAL SOLUBILITY AND PHASE FORMATION TENDENCY IN HEAs

In the very beginning of exploring HEAs in 1995, around 40 equiatomic compositions were tried. The alloy design was mainly based on the common elements (Huang and Yeh, 1996). The selection of elements avoided those such as Pb, Sn, and Bi which are chemically incompatible with some transition metals. Volatile elements, expensive rare elements and noble metals were also not considered. The equiatomic composition was used in each case as it would give maximum mixing entropy to enhance the mixing of constituent elements. Three categories of alloys, namely, Cu-containing, Al-containing, and Mo-containing were prepared using this approach. In the Cu-containing alloys, Co, Cr, and Pd were added into CuTiVFeNiZr alloy successively to generate septenary, octonary, and nonary alloys. Cu has weaker bonding with some transition metals and is even repulsive to Cr, Fe, and Co. In the Al-containing alloys, with similar additions to AlTiVFeNiZr alloy, Al could increase the bonding energy by its strong bonding with transition elements. In the Mo-containing alloys, similar additions to MoTiVFeNiZr alloy, Mo could increase Young's modulus because of its much higher modulus. From this first study, it led to a valuable observation that there is high-entropy effect which enhances the mixing of elements since no complicated structures were found. In the early publications, the discussion on the formation of phases were also based on atomic size difference, crystal structure, formation enthalpy of unlike atomic pairs, H-R rules for binary systems, and slower kinetics

(Ranganathan, 2003; Yeh et al., 2004b; Tong et al., 2005b; Yeh, 2006). By discussing the EDS compositions of phases and considering the atomic size, electronegativity, crystal structure, and chemical valence of all composing elements of $Al_x CoCrCuFeNi$ alloys ($x = 0-3.0$), Tong et al. (2005b) concluded that the high mixing entropy effect could enhance the formation of simple solid solution phases rather than intermetallic compounds and terminal solid solutions, and relax H-R rules. However, in the early development of HEAs no quantitative criteria and approaches existed to predict the solid solution formation in these multicomponent alloys that could lead to the best choice of elements in alloy design. Recent efforts utilizing more scientific criteria and approaches are dealt with in this section.

3.3 PARAMETRIC APPROACHES TO PREDICT CRYSTALLINE SOLID SOLUTION AND METALLIC GLASS

HEAs were developed much later than metallic glasses. HEAs and metallic glasses are all multicomponent. Furthermore, HEAs are quite close to bulk metallic glasses (BMGs) in highly concentrated compositions. The main difference is that BMGs are based on a single major component (>40 at.%). There have been some attempts to correlate the tendency of solid solution formation and glass forming ability (GFA). It is instructive to compare the criteria in predicting their formation.

3.3.1 Criteria for Glass Formation

The first attempt to rationalize metallic glass formation in alloys was by Turnbull (1969), a decade after the seminal discovery of metallic glass by Pol Duwez (Klement et al., 1960) in Au−Si alloy. Turnbull (1969) suggested a parameter, reduced glass transition temperature T_{rg} ($= T_g/T_l$), where T_g and T_l are glass transition temperature and liquidus temperature, respectively, as a measure of GFA of an alloy. Alloys with T_{rg} greater than 0.6 are identified as good glass formers. Among a number of other, ΔT_x ($= T_x - T_g$) (Inoue, 1995) (where T_x is crystallization temperature of glass), γ ($= T_x/(T_g + T_l)$) (Lu and Liu, 2002), α ($= T_x/T_l$), and β ($= (T_x/T_g) + (T_g/T_l)$) (Mondal and Murty, 2005) have become very popular. Figure 3.2 demonstrates the success of α and β parameters in identifying good glass forming compositions among about 50 different alloys. α parameter, in particular, has the advantage that it can be applied even to those glasses which do not show clear T_g as it is very close to T_x.

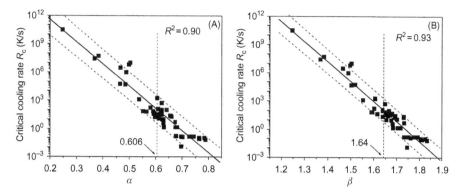

Figure 3.2 Influence of (A) α and (B) β parameters on critical cooling rate of about 50 different metallic glasses (Mondal and Murty, 2005).

Until late 80s glassy structure in metallic alloys was obtained only in some binary and ternary alloys, which did not possess high GFA. Hence these alloys needed to be rapidly solidified at cooling rates of the order of 10^6 K/s to induce glass formation, thus restricting the thickness of these metallic glasses to a few tens of microns. Inoue's group has pioneered research on BMGs from late 1980s that has led to the development of large sized glasses of up to 3" diameter with extremely high GFA (Inoue et al., 1989).

Metallic glasses with a critical thickness of larger than 1 mm are usually referred to as BMGs. In an attempt to identify the conditions required for an alloy to form a BMG, Inoue (1996) suggested three empirical rules. These are namely, (1) a multicomponent alloy with at least three components, (2) large negative enthalpy of mixing among the major constituent elements, and (3) atomic size difference larger than 12% among the main constituent elements. Among these the first criterion is related to kinetics suppression of crystal formation from liquid and it is usually referred to as "confusion principle" (Greer, 1993). The second criterion is related to the thermodynamic stabilization of the liquid, which is reflected in the form of deep eutectics in phase diagrams. The third criterion is related to topological aspect of formation of dense random packing, which is the nature of packing observed in glass. Based on these rules, Takeuchi and Inoue (2005) were able to classify the BMG alloys into seven groups consisting of five chemical types of elements (Figure 3.3). Interestingly out of these three rules, the last two are against the solid solution formation emphasized in HEAs.

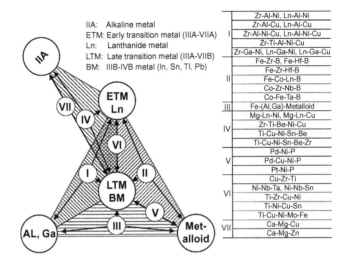

Figure 3.3 Five chemical types and seven groups of BMGs suggested by Inoue (Takeuchi and Inoue, 2005).

A large negative enthalpy of mixing ensures ordering and hence cannot stabilize random crystalline solid solution. Similarly, a large size factor is against the H-R rule for the solid solution formation and hence would destabilize a crystalline solid solution and enhance the GFA. The topological approach to glass formation is largely due to the work of Egami and Waseda (1984) and Miracle (Senkov and Miracle, 2001). Miracle (2006) has been able to identify glass forming composition ranges for a variety of classes of glasses based on the topological approach (Figure 3.4), where they suggest that a good glass forming system should have certain fractions of large-, small-, and intermediate-sized atoms.

Early thermodynamic efforts to evaluate GFA used enthalpy of mixing obtained from Miedema calculations, originally demonstrated in binary systems and later extended to multicomponent systems, to separate solid solutions from amorphous phase formation domains. Figure 3.5 shows such an effort in case of Ti−Ni−Cu system (Murty et al., 1992) for identifying ternary compositions that can lead to amorphous phase formation. These results were found to be close to those experimentally obtained by MA in this system.

3.3.2 Thermodynamic Parameters to Predict the Formation of Solid Solution and Metallic Glass

Thermodynamics parameters like enthalpy of mixing (ΔH_{mix}), entropy of mixing (ΔS_{mix}), and topological parameters like δ (Zhang et al., 2008b)

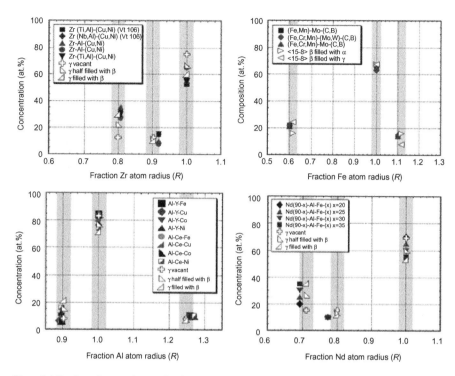

Figure 3.4 Topological approach to predict the composition ranges for BMG formation in different classes of multicomponent alloys (Miracle, 2006).

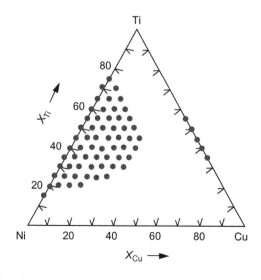

Figure 3.5 Miedema-based predictions for phase selection between amorphous phase and solid solution in Ti−Ni−Cu system (Murty et al., 1992).

and mismatch entropy (ΔS_σ) were used (Takeuchi et al., 2013a) to demarcate the BMG and solid solution formation conditions in multicomponent alloys. Zhang et al. (2008b) showed that simple solid solutions form in alloys with parameters satisfying the criteria: $-20 \leq \Delta H_{mix} \leq 5$ kJ/mol, $12 \leq \Delta S_{mix} \leq 17.5$ J/K mol, and $\delta \leq 6.4\%$ (Figure 3.6). When only disordered phases are allowed, the criteria are more restrictive, requiring a smaller negative limit of ΔH_{mix} (≥ -15 kJ/mol) and a smaller δ ($\leq 4.6\%$).

Here δ is defined as

$$\delta = 100 \sqrt{\sum_{i=1}^{n} c_i (1 - r_i/\bar{r})^2}$$

wherein c_i and r_i are composition and atomic radii of ith element and $\bar{r}$ is the average atomic radius.

Later, they used another thermodynamic parameter for prediction of solid solution formation as (Zhang et al., 2012c; Yang and Zhang, 2012):

$$\Omega = |T_m \Delta S_{mix} / \Delta H_{mix}|$$

where $T_m = \sum_{i=1}^{n} X_i (T_m)_i$ is the hypothetical melting temperature calculated according to the rule of mixtures. They placed a new criterion that simple solid solutions form when $\Omega \geq 1.1$ and $\delta \leq 6.6\%$ (Zhang et al., 2012c).

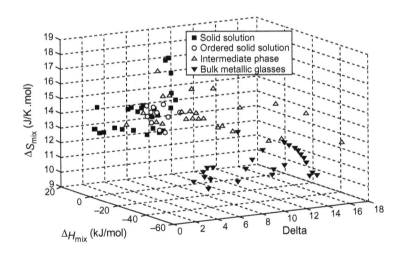

Figure 3.6 The effect of ΔS_{mix}, ΔH_{mix}, and δ on the phase formation in HEAs and typical multicomponent BMGs (Zhang et al., 2008b).

Guo and Liu (2011) pointed out that ΔH_{mix}, atomic size difference (δ) and mixing entropy (ΔS_{mix}) should all be used together to identify solid solution formation. They suggested that solid solution is expected if all the three parameters satisfy the conditions of $-22 \leq \Delta H_{\mathrm{mix}} \leq 7\,\mathrm{kJ/mol}$, $0 \leq \delta \leq 8.5$, and $11 \leq \Delta S_{\mathrm{mix}} \leq 19.5\,\mathrm{J/}$ $\mathrm{K\,mol}$. They identified that atomic size difference is the critical parameter for identifying solid solution and BMG forming HEAs. They suggested that BMGs form when $-49 \leq \Delta H_{\mathrm{mix}} \leq -5.5\,\mathrm{kJ/mol}$, $\delta \geq 9$, and $7 \leq \Delta S_{\mathrm{mix}} \leq 16\,\mathrm{J/K\,mol}$.

Guo et al. (2013a) used 2D maps of ΔH_{mix} and δ to identify the conditions for the formation of solid solutions, intermetallics, and amorphous phases in HEAs as shown in Figure 3.7. They pointed out that solid solutions can form when $\delta \leq 6.6$ and $-11.6 < \Delta H_{\mathrm{mix}} < 3.2\,\mathrm{kJ/mol}$. The amorphous phase can form when $\delta > 6.4$, and ΔH_{mix} is significantly negative ($\Delta H_{\mathrm{mix}} < -12.2\,\mathrm{kJ/mol}$). In a similar study, Ren et al. (2013) have identified the conditions for solid solution formation in HEAs as $\delta \leq 2.77$ and $\Delta H_{\mathrm{mix}} \geq -8.8\,\mathrm{kJ/mol}$. Although the above criteria have some inconsistence between them, they basically are useful guidelines to design compositions with disordered and/or ordered solid solutions.

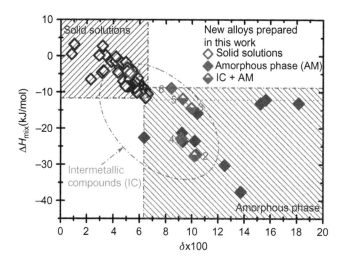

Figure 3.7 A $\delta - \Delta H_{mix}$ plot delineating the phase selection in HEAs. The dash-dotted regions highlight the individual region to form solid solutions, intermetallic compounds and amorphous phases (Guo et al., 2013a).

Wang et al. (2009a) and Sriharitha et al. (2013) observed a conversion from FCC structure to BCC when the Al content in AlCoCrCuFeNiTi and AlCoCrCuFeNi is increased. Wang et al. (2009a) attributed this to lattice distortion which destabilizes the close-packed FCC structure. Guo et al. (2014) have also observed that the solid solution decomposes when the lattice distortion crosses a value of 0.018 and this was observed when the number of elements increased to four and five when they studied Ni, CoNi, CoFeNi, CoCrFeNi, and CoCrCuFeNi. A close observation of their results suggests that the decomposition is actually observed only for the quinary alloy and the second FCC phase that is observed in this case is possibly due to the segregation of Cu, due to its positive ΔH_{mix} with Fe and Cr. Thus, it is important to have an unbiased study while identifying the conditions that lead to certain phase formation in HEAs. Singh and Subramaniam (2014) have also pointed this out recently based on a detailed assessment of the results of a number of ternary, quaternary, and quinary equiatomic alloys.

In contrast to the above observations, Otto et al. (2013b), based on the individual substitutions of different 3D and 4D transition elements to a quinary alloy (CoCrFeMnNi), found that it is the free energy minimization that decides the phase formation and neither ΔS_{mix} nor ΔH_{mix} can decide this independently. Thus, attempts to maximize configurational entropy alone do not lead to the formation of single-phase solid solutions and one needs to consider the interactions between the elements, attractive or repulsive, that can lead to intermetallics or phase separation. In addition, atomic size difference also can play a crucial role, particularly in deciding the phase formation between crystalline solid solution and amorphous phases.

As for mismatch entropy which concerns with the atomic size difference among composing elements, Takeuchi et al. (2013a) had calculated it through the following equations:

$$\frac{\Delta S_\sigma}{k} = \left[\frac{3}{2}(\zeta - 1)y_1 + \frac{3}{2}(\zeta-1)^2 y_2 - \left\{\frac{1}{2}(\zeta - 1)(\zeta - 3) + ln\zeta\right\}(1 - y_3)\right]$$

where

$$\zeta = \frac{1}{1-\rho}$$

ρ is packing fraction (0.64 for dense random packing). y_1, y_2, and y_3 are dimensionless parameters, expressed as

$$y_1 = \frac{1}{\sigma^3} \sum_{\substack{j>i=1}}^{n} (d_i + d_j)(d_i - d_j)^2 c_i c_j$$

$$y_2 = \frac{\sigma^2}{(\sigma^3)^2} \sum_{\substack{j>i=1}}^{n} (d_i d_j)(d_i - d_j)^2 c_i c_j$$

$$y_3 = \frac{(\sigma^2)^3}{(\sigma^3)^2}$$

where

$$\sigma^k = \sum_{i=1}^{n} c_i d_i^k \quad \text{(where } k = 2, 3)$$

where d_i is the atomic diameter of ith component.

The analysis of a large number of HEAs (Raghavan et al., 2012) indicates that solid solution formation in multicomponent alloys is favored when the ratio of $\Delta S_{conf}/\Delta S_{fusion}$ is greater than 1 and 1.2 for equiatomic and nonequiatomic alloys, respectively. The results also point out that BCC phase is favored when the atomic size difference is larger, which is reflected by a higher value of mismatch entropy ($\Delta S_\sigma/k$) as shown in Figure 3.8. FCC phase appears to form only when the

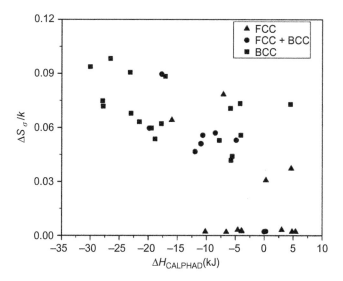

Figure 3.8 $\Delta S_\sigma/k$ versus $\Delta H_{mix(CALPHAD)}$ plot for equiatomic systems (Raghavan et al., 2012).

mismatch entropy ($\Delta S_\sigma/k$) is very small and the ΔH_{mix} is close to zero. This indicates that close-packed structures get stabilized when the system follows H-R rules and hence is close to ideal solution. In contrast, BCC phase gets stabilized when the mismatch entropy ($\Delta S_\sigma/k$) is very large and the ΔH_{mix} is more negative, indicating that open structures (BCC) can accommodate more strain and also nonideality.

Bhatt et al. (2007) have used the product of ΔH_{mix} and S_σ/k_B (P_{HS}) to effectively identify the compositions that are best glass formers in ternary and quaternary systems. However, this parameter was not found to be that effective in predicting the best glass forming composition in quinary and other higher order systems. However, when the configuration entropy is also considered in the form of a new parameter (P_{HSS}), which is a product of ΔH_{mix}, S_σ/k_B, and ΔS_{conf}, the prediction is found to be better. A more negative P_{HSS} value was found to lead to higher GFA in multicomponent systems (Rao et al., 2013).

Takeuchi et al. (2013a) statistically analyzed and compared the plots between ΔH_{mix} versus S_σ/k_B and ΔH_{mix} versus δ for 6150 ternary amorphous alloys from 351 systems. They found the relationship of $S_\sigma/k_B \sim (\delta/22)^2$ between mismatch entropy and atomic size difference. This again confirms that atomic size difference and mismatch entropy are basically correlated with each other. In the comparison of the plot for HEAs, BMGs and high-entropy BMGs, they found that new HEAs with simple solid solution may occur when -50 kJ/mol $\leq \Delta H_{mix} \leq -40$ kJ/mol, and $6 \leq \delta \leq 10$. The evidence is the $Al_{0.5}CuNiPdTiZr$ alloy with BCC single structure falling in this region. Thus, they provide a possible opportunity to design simple solid solution in difference regions.

Guo et al. (2011) found that the valence electron concentration (VEC) is the critical parameter for determination of BCC and FCC phase formation in HEAs. FCC phase forms when VEC ≥ 8 whereas BCC phase forms when VEC < 6.87. FCC and BCC coexist in between. This criterion is interesting since the FCC stabilizers or formers such as Co and Ni have VEC $= 9$ and 10, whereas BCC stabilizers such as Al and Ti have VEC $= 3$ and 4, respectively. That implies that more FCC elements tend to form FCC phase. Tsai et al. (2013c) have used this approach to delineate the σ phase formation domain based on the phases formed after aging of a number of HEAs (Figure 3.9). The side and central domains indicate the absence and presence of σ phase after aging, respectively. The aging condition for the $Al_xCrFe_{1.5}MnNi_{0.5}$ alloys and the test alloys is 700°C 20 h, while the rest of the alloys are collected from the literature.

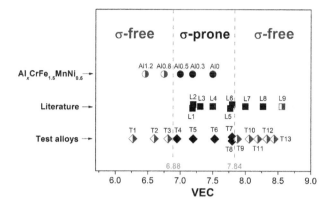

Figure 3.9 Conditions delineating σ phase formation in HEAs. Adapted from Tsai et al. (2013c).

This criterion could be combined with Guo's criterion to reveal the tendency of FCC and BCC phase to form σ phase during aging or annealing at intermediate temperatures between about 600°C and 1000°C.

3.4 PETTIFOR MAP APPROACH TO PREDICT THE FORMATION OF INTERMETALLIC COMPOUND, QUASICRYSTAL, AND GLASS

Alloying is a powerful method for changing the structure of the materials. The structure of alloys can be broadly categorized into four different types: crystalline solid solution, crystalline intermetallics, quasicrystalline intermetallics, and glass. The ability to predict the structure of a material is one of the fundamental challenges of materials research. The original H-R rules for the mutual solubility of two components include three factors: atomic size, electronegativity, and valency. Therefore, it is possible to predict the extent of solid solutions by considering these three factors. Pettifor (1984) has pioneered the structure of binary crystalline intermetallics. He has added a fourth factor of bond orbitals along with size, electronegativity and valency and has created a chemical scale. In this chemical scale, each element in the periodic table is assigned a unique number called the Mendeleev number. The Mendeleev number is a phenomenological coordinate used by Pettifor, which is determined by running a 1D string through the 2D periodic table (Figure 3.10 and 3.11). The path of the string is chosen in such a way to give best structural separation of AB compounds. Therefore, elements which are neighbors in their atomic numbers could have very different Mendeleev numbers and elements with different atomic numbers could be neighbors with respect to their

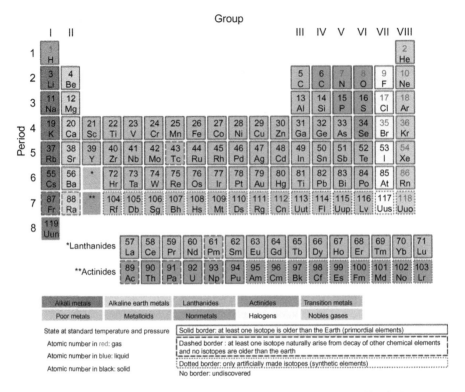

Figure 3.10 Periodic table of 118 elements (http://commons.wikimedia.org/wiki/Periodic_Table_of_Elements).

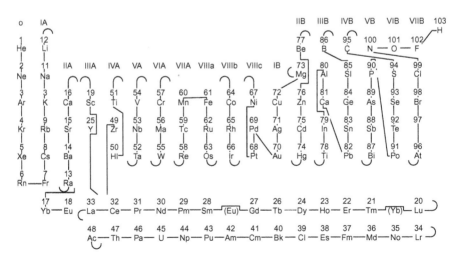

Figure 3.11 Pettifor assignment of Mendeleev numbers to elements using a string through the modified periodic table (Ranganathan and Inoue, 2006).

Mendeleev numbers. This single number can be used to differentiate various structures.

Based on this concept a map is constructed and boundaries are drawn to separate compounds of different structure types. Bare patches in the map show the regions of positive heats of formation, which correspond to binary nonformer. Pettifor (1988) has also shown that the structure mapping approach can be used to predict the structure of ternary and quaternary compounds by treating them as pseudobinaries (Pettifor, 1988). He has used the method of calculating average Mendeleev number for this prediction. If A_xB_y is considered as the binary alloy and ternary and quaternary additions C and D preferentially go the A and B sites, respectively, then the alloy can be treated as a pseudobinary alloy $(A_xC_{1-x}) (B_yD_{1-y})$. The pseudobinary alloy is characterized by the average Mendeleev numbers M_{*A} and M_{*B} which are given by $M_{*A} = xM_A + (1 - x)M_c$ and $M_{*B} = yM_B + (1 - y)M_D$. This simple scheme shows that the structural domains of pure AB and AB_3 binaries are similar to that of the pseudobinaries.

Nearly 20 years after the advocacy of this structure mapping approach, Villars et al. (2001, 2004) used the Mendeleev number for the prediction of former/nonformer in any binary, ternary, and quaternary systems (Figure 3.12). They reported that the Mendeleev number is a highly effective parameter as compared to the atomic number for this structure prediction method.

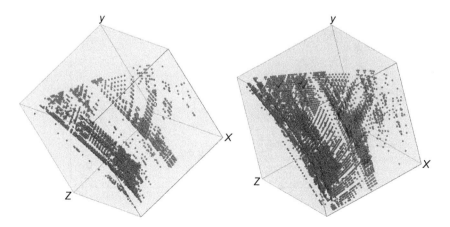

Figure 3.12 Prediction of binary, ternary, and quaternary compound former/nonformer alloys using Mendeleev number (Villars et al., 2001).

Jeevan and Ranganathan (2004) used the Pettifor pseudobinary structure mapping approach for the classification of quasicrystals. They have shown that AB_6 and $A_{13}B_5$ compounds are related to quasicrystals and are actually the extension of the AB and AB_3 compounds. They were able to understand the quasicrystal forming ability of various compositions in Zn–Mg–RE and Cd–Mg–RE systems using this approach. Ranganathan and Inoue (2006) have used this approach for the identification of higher order quasicrystal forming alloys (ternary and quaternary) as pseudobinary quasicrystalline intermetallics. They considered the largest sized atom as the most important constituent. They have been also able to classify quasicrystals into four structural classes (Bergman class, Mackey class, Kuo class, and Tsai class) based on the nature of the bond orbital s, p, d, or f of the large atom with four associated related crystal structures. They have also identified that atomic size together with Mendeleev number can give a better systematic classification of quasicrystals to different groups (Figure 3.13). They observed that Mendeleev number is the most important parameter in determining the quasicrystal forming ability of alloys. They also pointed out that atomic size factor is the dominant parameter for quasicrystalline alloys rather than composition. They argued that a minimum amount of the large-sized atom is necessary to create icosahedral order in the phase. They have shown that Pettifor's average Mendeleev number scheme is also applicable to A_2B_3, A_5B_2, AB_2, and AB_6 stoichiometries and related quasicrystals.

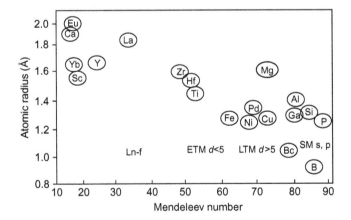

Figure 3.13 Atomic size versus Mendeleev number for the elements that are important for quasicrystal and BMG formation (Ranganathan and Inoue, 2006).

Biswas and Ranganathan (2006) suggested, based on Mendeleev number that many higher order alloys can be visualized as pseudo-lower-order alloys. Disordered solid solutions can be seen to be pseudounary, crystalline intermetallics can be at the highest level pseudobinary. They also demonstrated that most quasicrystals and their rational approximants are pseudobinary and that most metallic glasses are pseudoternary. They were able to demonstrate correlation between the Mendeleev number and GFA in Mg–Cu–Re alloys. The GFA appears to reach a maximum at a Mendeleev number of 27 (Gd). They also observed that Mg–Cu–Gd alloys have better GFA in comparison to Mg–Cu–Nd, due to the higher Mendeleev number of Gd compared to Nd.

Takeuchi and Inoue (2006) were the first to use this approach for the classification of BMGs. They have adopted Pettifor's pseudobinary structure mapping approach to analyze the characteristic of atomic pairs in ferrous BMGs. They have described the multicomponent BMG forming systems as the sum of pseudobinary systems. For example, a ternary $A_a B_b C_c$ (A, B, and C are the constituent elements and a, b, and c are the atom percentages of the respective elements) BMG can be described as the sum of three pseudobinaries: $A_{a/(a+b)} B_{b/(a+b)}$, $B_{b/(b+c)} C_{c/(b+c)}$, and $C_{c/(c+a)} A_{a/(c+a)}$. Multicomponent BMG systems also can be described as the sum of many pseudobinaries.

Takeuchi et al. (2007) have analyzed BMGs with a tetrahedron composition diagram that comprises constituent classes from blocks of elements in the periodic table. Figure 3.14 shows the grouping of BMGs based on the classes of constituents, namely, s-, d_Ef-, d_Lp-, and p-blocks. When the BMGs contain Al and Ga, they are assumed to correspond to either s- or p-block elements. This analysis revealed the presence of a composition band connecting different classes of BMGs (Figure 3.15). They could show that this diagram is applicable to any multicomponent alloy system and glass forming compositions can be analyzed from the bonding nature of the atomic pairs. In addition, they indicated that such composition diagrams can be used for analyzing other metallic materials such as quasicrystalline alloys and intermetallic compounds.

Curtarolo et al. (2013) carried out high-throughput analysis of binary intermetallics and were able to segregate compound forming and noncompound forming systems based on Mendeleev number using Pettifor map approach (Figure 3.16).

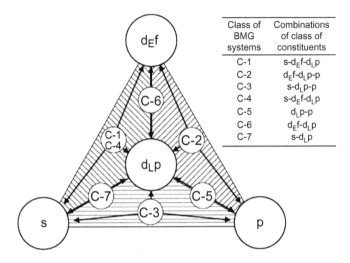

Class of BMG systems	Combinations of class of constituents
C-1	s-d_Ef-d_Lp
C-2	d_Ef-d_Lp-p
C-3	s-d_Lp-p
C-4	s-d_Ef-d_Lp
C-5	d_Lp-p
C-6	d_Ef-d_Lp
C-7	s-d_Lp

Figure 3.14 Classification of BMGs based on the classes of constituent elements as the s-block elements (s), early transition metals and f-block elements (d_Ef), late transition and p-block metals (d_Lp) and metalloids (p) (Takeuchi et al., 2007).

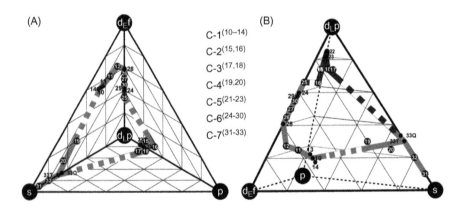

C-1$^{(10–14)}$
C-2$^{(15,16)}$
C-3$^{(17,18)}$
C-4$^{(19,20)}$
C-5$^{(21-23)}$
C-6$^{(24-30)}$
C-7$^{(31-33)}$

Figure 3.15 s-d_Ef-d_Lp-p composition diagram; (A) in its plane projection and (B) 3D drawing with plots of typical BMGs from the seven classes of the BMG (Takeuchi et al., 2007).

3.5 PHASE SEPARATION APPROACH TO FIND SINGLE-PHASE HEAs

From the different approaches mentioned above, it can be said that there are still no universal criteria to predict the formation of different phase types. Even so, these approaches and predictions are very helpful for composition design with an aim to achieve required phase types such as disordered solid solution, partially ordered solid solution, intermetallics, quasicrystalline compounds, and metallic glasses.

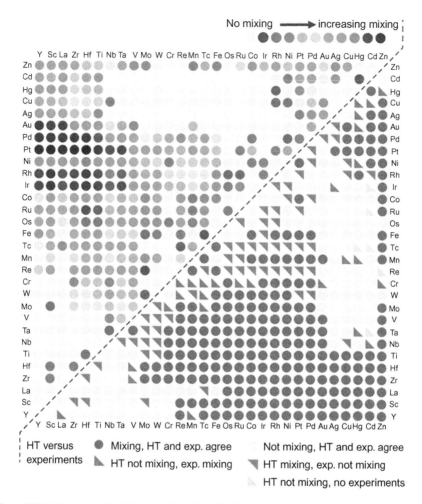

Figure 3.16 Pettifor map to identify the conditions for solid solution and compound formation in different binary systems (Curtarolo et al., 2013).

In addition to the above principles for judging the phase type of constituent phases of a HEA composition, one might directly obtain the compositions for generating monolithic-phase HEAs from the EDS (energy dispersive spectrometer) or EPMA (electron probe microanalyzer) analyses on each major phase of a multiphase alloys. This method has often been used in the development of conventional alloys. Suppose a binary alloy has two phases, α and β, in the microstructure. Then, we can use the composition of α phase to prepare simple α-phase alloy. For a more conservative way, the β-phase forming elements might be reduced suitably to let the composition safely fall into α-phase region. Similarly, simple β-phase alloy could be obtained. This methodology could be equally applied to HEAs for creating more single-phase alloys, single matrix alloys with a dispersion of second phases, or simple spinodally decomposed alloys. Obviously, these compositions are generally nonequiatomic.

Depending on the property required for a particular application, a particular phase in the multiphase alloy is targeted. For example, it has been well known that Cu in the alloy system $Al_xCoCrCuFeNi$ tends to segregate in the interdendritic region due to its significant repulsion to Cr, Fe, and Co. Tables 3.1 and 3.2 show the chemical composition of dendrite and interdendrite, respectively, for $Al_{0.5}CoCrCuFeNi$ alloy in different states (Tsai et al., 2009b). From the dendrite composition, new single FCC phase alloy at high temperatures might be designed. Similarly, interdendrite composition might provide new concentrated Cu alloys.

It should be mentioned that by Gibbs phase rule, the degree of freedom for a single phase in composition is high. So, one can further adjust the concentration of each element to improve some properties but still maintain the single phase. In fact, most HEA researchers have used such principles to develop new HEAs for desired properties from the beginning. In general, equiatomic HEAs are the beginning points for understanding new alloy systems. Based on this, nonequiatomic HEAs were derived for better performance. Then, fine tuning of composition and process were also done in order to achieve aimed properties just like the historical development of high-performance alloys.

Table 3.1 Composition of Dendrite Regions in $Al_{0.5}CoCrCuFeNi$ Alloy Obtained by EDS Analysis (in at.%) (Tsai et al., 2009b)

State	Al	Co	Cr	Cu	Fe	Ni
As-cast	9	19	21	11	20	20
As-homogenized-FC	7	21	22	11	21	18
As-homogenized-WQ	7	20	21	11	22	18

Table 3.2 Compositions of Interdendrite Regions in $Al_{0.5}CoCrCuFeNi$ Alloy Obtained by EDS Analysis (in at.%) (Tsai et al., 2009b)

State	Al	Co	Cr	Cu	Fe	Ni
As-cast	12	5	4	61	5	13
As-homogenized-FC	7	3	2	75	3	11
As-homogenized-WQ	10	5	4	59	6	15

CHAPTER 4

Alloy Design in the Twenty-First Century: ICME and Materials Genome Strategies

4.1 INTRODUCTION

Lord Kelvin stated that "To understand, you must be able to measure it." John von Neumann asserted that "To understand, you must be able to compute." In a lecture, Richard Feynman stated that "To understand you must be able to create it." There has been a spectacular progress in biology along these directions in the quest to understand life. The triumph of biology is the double-helix, genetic code and the recent synthesis of an artificial cell. Is there a corresponding Materials Genome?

Materials scientists are ready to grasp the challenge posed by J.C. Slater in 1956: "I don't understand why you metallurgists are so busy in working out experimentally the constitution (crystal structure and phase diagram) of multinary systems. We know the structure of the atoms (needing only the atomic number), we have the laws of quantum mechanics, and we have electronic calculation machines, which can solve the pertinent equation rather quickly."

Recently, Integrated Computational Materials Engineering (ICME, 2008) has become an emerging multidisciplinary field that is concerned with an approach to design products, the materials that comprise them and their associated processing methods by linking materials models at multiple length scales. The key links are just the cores of Materials Science, that is, process—structures—properties—performance (Figure 4.1). The need to understand different scales of structure in materials has been brought out long back by Smith (1981) (Figure 4.2). The importance of multiscale modeling has been brought out very elegantly by Olson (1997, 2013), who could elucidate the hierarchy of computational models and the characterization tools that need to be used to validate the models while designing the materials (Figures 4.3).

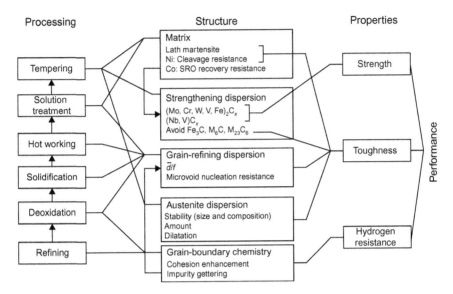

Figure 4.1 The correlation of processing–microstructure–properties–performance for an advanced alloy steel (Olson, 2013).

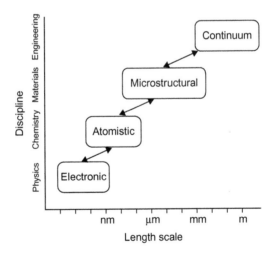

Figure 4.2 The various levels of structure in materials.

ICME uses a variety of simulation software tools in combination to accelerate materials development and unify design and manufacturing. Although major developments have successfully been driven essentially by academic and industrial users, ICME is still in its infancy.

Hierarchy of design models

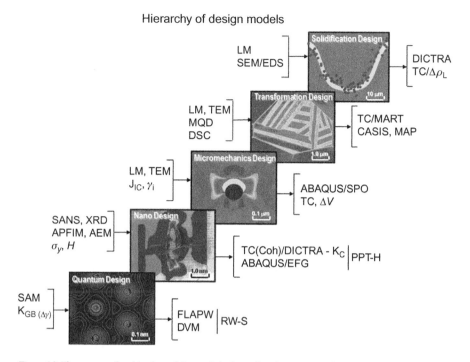

Figure 4.3 The concept of multiscale modeling and the hierarchy of computational design models and the experimental tools that are used to validate the models (Olson, 2013).

Advanced materials are essential to address challenges in clean energy, national security, and human welfare, but it can take 20 or more years to realize a new material in the market. Therefore, accelerating the discovery and realization of advanced materials will be crucial to achieving global competitiveness in the twenty-first century. Based on the promising capability of ICME as reported in the 2008 study published by the National Research Council (ICME, 2008), ICME together with Accelerated Technology Transition (ATT) study announced by National Research Council in 2004 has paved way for rapid strides for computational materials science.

USA launched an ambitious plan, the Materials Genome Initiative (MGI), to double the speed with which we discover, develop, and manufacture new materials in June 2011 (National Science and Technology Council of US, 2011). This call has caught the imagination of the world scientific community. Figure 4.4 (Drosback, 2014) shows the various ingredients that are required to realize this ambitious initiative. The program will fund computational tools, software, new methods

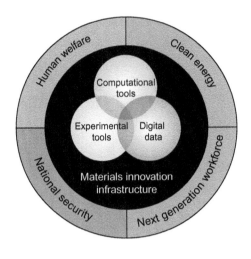

Figure 4.4 An overview of the MGI, showing the key components of the infrastructure and areas of application. Courtesy: National Science and Technology Council of US (2011).

for material characterization, and the development of open standards and databases that will make the process of discovery and development of advanced materials faster, less expensive, and more predictable. The recent view point set published in *Scripta Materialia* (Olson, 2014) reiterates the importance of the MGI. In this view point set, Olson and Kuehmann (2014) bring out clearly the time line for MGI. The statement of Olson and Kuehmann (2014) "Moving beyond a system of 'technology by accident' grounded in near-random-walk scientific discovery to a new system of tightly integrated science-based engineering constitutes a true revolution in materials technology, offering substantial societal benefit as well as a new justification for scientific investment. Clear opportunities exist for the further expansion of the existing Materials Genome system to further enhance its value creation potential" brings out the scope and future of MGI.

As high-entropy alloys (HEAs) provide numerous combinations of new compositions and properties, and thus potential applications for human benefit, effective alloy design including composition and process selection is especially important. Chapter 3 has explained the phase prediction from mixing entropy, mixing enthalpy, atomic size difference, and valence electron concentration. While this is helpful in some aspects of alloy design, more quantitative predictions are still far from sufficiency and require lots of experiments to understand them.

As a result, newly developed ICME becomes very important to identify and realize new HEA materials in an effective and timely way. Although the database is still lacking for such multicomponent materials, it is nevertheless the effective way to explore and exploit them in the future.

4.2 INTEGRATED COMPUTATIONAL MATERIALS ENGINEERING

The name ICME suggests an "integration" of a number of computational models applicable at different length scales to "materials engineering" which is involved in answering questions regarding the identification and creation of novel materials for the wide-scale applications of the twenty-first century.

Within ICME, the field of CALPHAD method, which involves the collection of thermodynamic databases, determination of phase diagrams pertaining to the thermodynamic information in the databases, and the use of the computational techniques that can be utilized for the calculations of thermodynamic properties regarding phase coexistence and phase stability, is thought to be very important for MGI. More elaborately, the CALPHAD methodology can be used to determine relative phase stabilities, individual component solubilities, transition temperatures, rate constants, and phase fractions. Clearly, such knowledge enables the choice of not only the materials required for a given application, but also the compositions applicable for the synthesis of these materials, which is a crucial component of ICME.

Besides the CALPHAD method, computational models developed for different length scales also contribute to calculate, explain, and predict the pertinent behavior and properties of a material. These simulation techniques include *ab-initio* methods such as density functional theory (DFT), atomistic methods such as Monte Carlo (MC) and molecular dynamics (MD), and continuum techniques such as phase field and finite element. While simulation techniques such as DFT are usually used in order to augment the information already existing in databases or create new databases for calculated phases, the other techniques (MD and MC) are utilized for answering questions regarding structures of nucleating phases/phase clusters and initial growth using information obtained through fitting to experimental data or *ab-initio* DFT computations. Phase-field and finite element methods are involved in investigating questions of microstructure evolution and pattern formation occurring during materials processing, which also utilize thermodynamic information from databases. Indeed, combining and utilizing these computational methods provides a methodology for creating (process–structure) and (structure–property) correlations, thereby accelerating the pace of discovery and development of materials. In this chapter, this conglomeration of methods will be individually discussed in the context of application to HEAs.

4.2.1 CALPHAD Method

About 35 years ago, Larry Kaufman and Himo Ansara brought together a small number of scientists working on the calculation of alloy phase diagrams. This has been the origin of CALPHAD (CALculation of PHAse Diagrams) and computer coupling of phase diagrams and thermochemistry (Kaufman and Cohen, 1956). CALPHAD has become a successful and widely applied tool in many areas of materials development (Spencer, 2008). Kaufman and Agren (2014) in a recent article bring out the road map for quantitative microstructure engineering through CALPHAD (Figure 4.5), which includes atomistic calculations using methods such as DFT. DFT is a quantum mechanical modeling technique used to calculate the electronic structure in single atoms and atom clusters/molecules and condensed phases. In this technique, the properties of a many-electron system can be determined using functions of electron density, which are referred to as functionals. The name "DFT" comes from the use of functionals of the electron density. The solutions of the electron densities represented as wave functions and their associated energies are derived from the Schrodinger equation, usually utilizing one or more simplifications to reduce complexity. DFT is one of the most widely used methods in computational physics and chemistry.

Boesch and Slaney (1964) proposed the prediction of existence of harmful sigma phase in Ni-base superalloy by calculating average electron hole number and judging whether it exceeds the critical value for sigma phase formation. This is called PHACOMP (abbreviation of PHAse COMPosition). After this, a new Phacomp was invented in 1984 using d-electron concept to define the phase boundaries in terms of M_d (metal d-level), especially define the critical M_d value of gamma phase and sigma phase boundary for predicting sigma phase formation in Ni-base superalloys (Morinaga et al., 1985). Although these two methods are simple and quick ways to see the degree to which an alloy

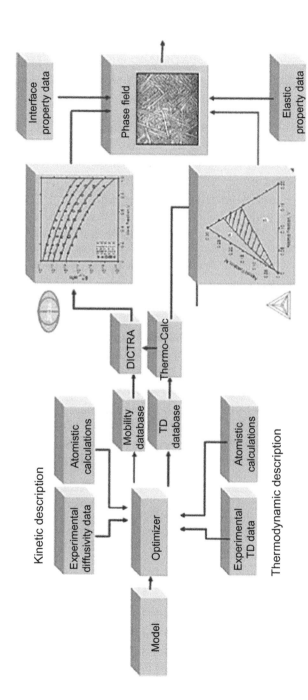

Figure 4.5 The road map for microstructure engineering using CALPHAD (Kaufman and Agren, 2014).

is prone to form sigma phase, they cannot predict other deleterious phases such as mu and Laves phases. Moreover, it cannot provide the temperature range of stability and phase boundaries. CALPHAD development has overcome these drawbacks for designing and developing superalloys by providing phase equilibrium, liquidus line, solidus line, transition temperature, phase amount, phase chemistry, etc.

CALPHAD can justifiably claim to be one of the major success stories in the field of materials development over the last quarter century. CALPHAD started in the 1960s but sophisticated thermodynamic data bank systems started to appear in the 1980s. In the CALPHAD method, firstly, available thermodynamic properties and phase equilibrium data of a system's constituents are collected. Then, a thermodynamic description of the system is obtained. This description is basically a mathematical model which can be used not only to reproduce the known thermodynamic information, but more importantly, it is hoped that one can predict unknown thermodynamic properties of the system.

The CALPHAD technique makes use of the principle that the Gibbs energy of a phase that is a function of temperature and composition is enough to obtain a complete thermodynamic description of the system since almost all thermodynamic properties can be derived from the Gibbs energy function (Muggianu et al., 1975; Palumbo and Battezzati, 2008). The entire thermodynamic description is contained in a thermodynamic data base (TDB) file, which contains the Gibbs energy functions of the respective phases, constructed out of suitable "solution" models. The starting point is to write the Gibbs energy of a phase φ as,

$$G^{\varphi} = {}^{\text{ref}}G^{\varphi} + {}^{\text{id}}G^{\varphi} + {}^{\text{ex}}G^{\varphi}$$

The first term on the right hand side of the above equation is contribution of pure components of the phase, the second is the ideal mixing contribution, and the third is the excess Gibbs energy of mixing. The second term is given by

$$^{\text{id}}G^{\varphi} = RT \sum_{i} x_i \ln x_i$$

where R is the universal gas constant, T is the temperature in absolute scale, and x_i is the mole fraction of the ith component. The third term is given by

$$^{\text{ex}}G^{\varphi} = \sum_{i} \sum_{j>i} x_i x_j \sum_{v} {}^{v}L_{ij}^{\varphi}(x_i - x_j)^{v}$$

where x_i and x_j are mole fractions of ith and jth components, respectively, and $^vL_{ij}^{\varphi}$ are model parameters (Reidlich–Kister parameters) that are found out as best fit for available data by using statistical techniques (Durga et al., 2012). Thus, thermodynamic description of any phase can be calculated.

This information is generally read in by an optimizing engine (e.g., Thermo-calc), which uses the thermodynamic information from the databases (TDB file) to derive the required thermodynamic properties and phase diagrams. In this technique, the stable/metastable states of the system are arrived through Gibbs energy minimization. The equations which result from the optimization procedure are a set of nonlinear equations involving the Gibbs energies of the corresponding phases and the unknown compositions, which are solved through optimization routines that are part of the software engine (e.g., Thermo-calc and Pandat). The results of the calculations are the phase-coexistence lines/surfaces/hyperplanes, representing the phase diagram. At present, there are several commercial products, for example, FactSage, MTDATA, PANDAT, MatCalc, JMatPro, and Thermo-Calc, which serve as intermediate engines for reading in thermodynamic databases, and deriving useful thermodynamic information about material properties and their response to some types of materials processing such as solidification. These resources serve as useful tools in research and industrial development for saving time and experimental work.

In case of multicomponent systems, this approach has a profound influence because thermodynamic descriptions of the constituents (binaries and ternaries) can be combined and extrapolated using geometric models to develop a thermodynamic description of a multicomponent system, which is otherwise either experimentally not possible or requires stupendous number of experiments. One must note, however, that created databases must be experimentally validated to establish confidence in the derived information.

Zhang et al. (2012a) used this approach in as-cast and homogenized alloys of Al–Co–Cr–Fe–Ni system and came up with the phase diagram for the system which confirms with experimental observations fairly well. In several multicomponent systems, the stable phases, their relative amounts, and individual compositions were predicted using the CALPHAD approach with appreciable degree of success in agreement with experimental results (Durga et al., 2012; Raghavan et al., 2012).

In another research (Durga et al., 2012), among the 22 multicomponent systems (quaternary and above) studied in the Co−Cr−Cu−Fe−Mn−Ni system, 14 of them show the FCC phase (Ni−Mn rich) to be the predominant phase, while 6 of them show the BCC phase (Co−Fe rich phase) to be the predominant one.

An attempt has been made to predict phase formation using a CALPHAD approach for a large number of equiatomic and nonequiatomic HEA compositions that are known to form FCC, BCC, and a mixture of FCC and BCC phases (Raghavan et al., 2012). The stable phase is assumed to be the first phase that is formed upon cooling from liquid state with the highest driving force. The driving force for other phases at the transition for various compositions is also calculated. The results indicate that solid solution formation in multicomponent alloys is favored when the ratio of $\Delta S_{conf}/\Delta S_{fusion}$ is greater than 1 and 1.2 for equiatomic and nonequiatomic alloys, respectively. CALPHAD approach appears to predict BCC phase formation much more accurately than the FCC phase formation. It is believed that this may be because of a greater presence of kinetic effects than in BCC, which is a more open structure. The results also point out that BCC phase is favored when the atomic size difference is larger, which is reflected by a higher value of mismatch entropy ($\Delta S_\sigma/k$). Formation of FCC phase appears to form only when the mismatch entropy ($\Delta S_\sigma/k$) is very small and the ΔH_{mix} is close to zero. This indicates that close-packed structures get stabilized when the system follows Hume-Rothery rules and hence is close to ideal solution. In contrast, BCC phase gets stabilized when the mismatch entropy ($\Delta S_\sigma/k$) is large and the ΔH_{mix} is more negative, indicating that open structures (BCC) can accommodate more strain and also nonideality.

Manzoni et al. (2013a) reported on two alloy systems of same elements with significantly different compositions and properties (brittle $Al_{23}Co_{15}Cr_{23}Cu_8Fe_{15}Ni_{16}$ and ductile $Al_8Co_{17}Cr_{17}Cu_8Fe_{17}Ni_{33}$) that the phase prediction by CALPHAD method was successful in the first alloy and only partially successful in the second alloy. Senkov et al. (2013b) have done CALPHAD analysis on high-hardness and low-density refractory multicomponent alloys of the Cr−Nb−Ti−V−Zr system. The predicted equilibrium room temperature phases with their volume fractions were found experimentally only at higher temperatures (600−750°C). This indicates that still slower cooling rate (<10°C/min) from the homogenization temperature of 1200°C may be necessary to achieve the predicted structure or there is a scope for better model.

4.2.2 *Ab-Initio* Calculations

Ab-initio calculations such as DFT involve the direct calculation of the electronic structure of atoms through the solution of Schrodinger equations (normally simplified with assumptions).

The determination of the electronic structure serves as a powerful tool to predict materials behavior and properties, which also involves the interaction between elements. Thus, one can determine formation energies, magnetic states, and lattice parameters for binary or multinary phases. This method augments the CALPHAD method when information regarding properties such as crystal structure are not available or the literature lacks sufficient experimental data. The downside, however, is that *ab-initio* calculations require a large amount of numerical computation, with the computing time rapidly increasing with the number of atoms.

Li et al. (2008b) used MD simulations and *ab-initio* calculations to find out the type of BCC phase formed with variation in Al content in AlCrCoNiFe HEA. In this study, FORCITE OF MATERIALS STUDIO and CASTEP computer software packages were used. These computer simulations and experiments agree that due to large electronegativity difference, compound formation tendency increases in this HEA system and results in the formation of ordered phase. For example when Al and Cr were used in higher amounts (Al + Cr ≥ 50 at.%), the solidified solid solution would have an ordered (B2) structure because Al and Cr have larger electronegativity difference. Therefore, it is suggested that while designing the alloy, higher amounts of Al and Cr can be avoided for better ductility, because B2 is a structure that is relatively brittle in nature.

Wang and Ye (2011) used first principle calculations to find out how lattice parameter and formation enthalpy are dependent on elemental composition in FeNiCrCuCo alloy. In this study, plane-wave pseudopotentials and alchemical pseudoatom methods were combined and used. It is demonstrated to be an efficient and reliable method to imitate the elemental positions in the lattice. In alloy design perspective, this study throws light on the role of Cr. Increase in formation enthalpy and decrease in lattice parameter occur with increasing Cr content. Such influence is attributed to electronic configuration and ionic radius of the element. Copper also is found to increase the formation enthalpy, but the reason is not investigated in this study.

Egami et al. (2013) reported a computer simulation study on the effect of irradiation on HEAs. Particle irradiation gives rise to atomic displacements and thermal spikes. Amorphization occurs easily in HEAs during irradiation. It occurs by atomic displacements resulting from irradiation and inherent high distortion at the atomic level of the HEAs. Following this, local melting and recrystallization occur due to thermal spikes. It is speculated by Egami et al. (2013) that such response to irradiation reduces defects in HEAs, and thus, HEAs are excellent candidates for nuclear application. Initial results of computer simulation on modeling binary alloys and an electron microscopy study on Hf−Nb−Zr alloys, demonstrate extremely high irradiation resistance of these alloys against electron damage to support this speculation.

Tian et al. (2013b) used *ab-initio* calculations and DFT for CoCrFeNiTi alloy system. Exact muffin-tin orbitals (EMTO) method in combination with the coherent potential approximation (CPA) is used. Accuracy of the single-site mean-field approximation is evaluated by comparing the CPA results with those generated by the supercell technique. Elastic modulus, equilibrium volume, atomic radius, and magnetic moments were estimated. Effect of variation of properties with amount of Ti was studied. Ti was found to increase anisotropy and ductility. But, calculations in this study seem to deviate significantly with experiments in several alloys indicating the scope for more extensive calculations and experiments.

One of the basic criteria to stabilize single-phase solid solution microstructure is the high-entropy effect. It arises essentially due to equiatomic composition and large number of elements. But, restricting the design only to equiatomic composition causes several constraints. Therefore, the range within which the amount of one element can be varied without changing the single-phase solid solution structure is of profound design importance. In this regard, a theoretical study using Bozzolo−Ferrante−Smith (BFS) method of atomistic modeling is reported (Del Grosso et al., 2012). By this, the lower concentration limit of each element in the alloy system with W, Nb, Mo, Ta, and V, to sustain single-phase solid solution, is estimated (Figure 4.6). It was mentioned in Chapters 2 and 3 that the effect of configurational

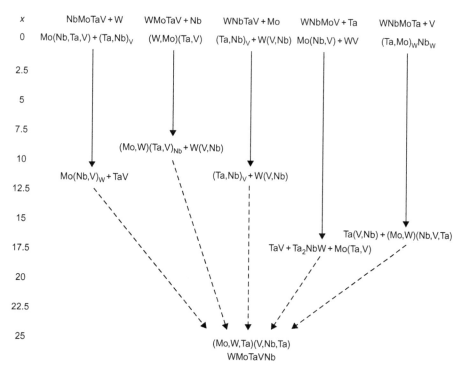

Figure 4.6 *Evolution of the different quaternary alloys to* W−Nb−Mo−Ta−V. *Each column indicates the phase structure corresponding to the concentration of the fifth element (*x, *in at.%) (Del Grosso et al., 2012).*

entropy is higher in stabilizing a single phase as the temperature increases. Thus, for some alloy systems there can be a lower temperature limit (critical temperature), above which the existence of solid solution is ensured. It is not difficult to perceive that finding such a temperature limit is of major help in alloy design. Estimation of this temperature is done as a function of amount of each element in the alloy system with W, Nb, Mo, Ta, and V.

4.2.3 MD Simulation

MD simulations are atomistic techniques, which are utilized usually for prediction of structures of initial cluster formation or the material response to stimuli right-down to the scale of atoms. An important input for MD simulation is the description of pair potentials which describe the interaction forces of the different atoms in the system as a function of separation. The equations of motion of each of these atoms is then derived from simple Newtonian mechanics where the total force on an atom is arrived through the super-position of the atomic forces from all other atoms, wherein the force between two atoms is derived using the described atomic potentials. Quite clearly, the predictive capability of this method lies in the degree of detail through which the pair potentials are constructed.

An exemplary application is in using a many-body tight-binding potential model to study the effect of number of elements and size difference on the amorphous structure of HEAs (Kao et al., 2006, 2008). This is simply because this model treats the interatomic forces existing between any two unlike atoms as the geometric average of their bonding forces in their respective pure lattices, and thus treats the systems as ideal solutions (i.e., mixing enthalpy is zero). In other words, such a MD simulation rules out the effect of actual bonding energy between unlike atoms, and only investigates the effects from the number of elements and the atomic size. The alloys simulated were from traditional binary alloys to HEAs by adding one element in sequence. For example, Figure 4.7 shows the initial radial distribution function curves of the alloys at 300 K before the system was heated. Thus, we can see that the patterns of binary to quaternary alloys have well-defined peaks which indicate a crystalline structure. However, quinary alloy and sexinary alloy containing large-sized Zr have lower and broader peaks which confirm an amorphous structure. Virtually, this trend apparently shows that the amorphization is enhanced by the number of elements and large atomic size difference.

When heated up to the molten state (2200 K) the patterns typically depict a liquid structure as shown in Figure 4.8A. Moreover, as the

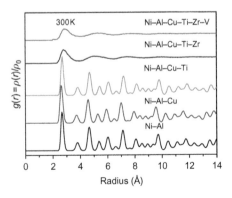

Figure 4.7 Radial distribution functions obtained at 300 K for equiatomic alloys from Ni–Al to Ni–Al–Cu–Ti–Zr–V by MD simulation.

number of elements increases, peaks become broader and distance between peaks also become larger which indicates that the liquid structure becomes more disordered. In the melt-quenched state, quinary and sexinary alloys exhibit liquid-like solid structure as shown in Figure 4.8B. But, binary to quaternary alloys display an amorphous structure because there is a splitting in their second peaks, which indicates that the structure is more ordered than the liquid structure. This again shows an increased number of elements, and thus large atomic size difference could enhance amorphization. In fact, the shape and evolution of radial distribution function curves could be explained from the close-packed hard ball model as shown in Figure 4.9. The splitting of the second peaks indicates that the second nearest neighbor shell is not fully merged with the third shell due to the insufficiency in the degree of disorder. By the hard ball model as shown in Figure 4.9, the number of atoms and radius for each shell are shown in the second and third rows of Table 4.1, respectively. Under a random occupation of sites by different atoms, the fluctuation range caused by the atomic size difference can be used to judge the merger of peaks. If the atomic size difference makes the atomic fluctuation range in the second and third shells larger than 7.2%, the second and third shells or peaks are expected to merge into each other. It can be seen from Table 4.2 that only quirary and sexinary alloys with deviation over 10% can fit this requirement. Since the deviation between fourth and fifth shell is 6.2%, all the alloys can have the merger of fourth and fifth peaks. Therefore, the judgment of peak merger by the atomic size difference is consistent with the radial distribution function calculated by MD simulation.

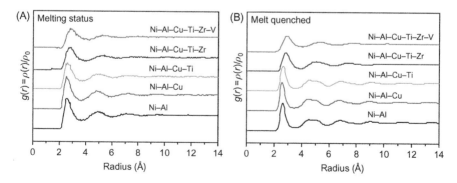

Figure 4.8 Radial distribution function obtained (A) at the melting state of 2200 K and (B) at the quenched state for equiatomic alloys from Ni−Al to Ni−Al−Cu−Ti−Zr−V by MD simulation. Adapted from Kao et al. (2006).

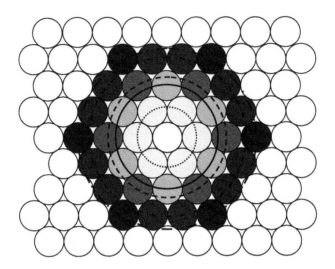

Figure 4.9 Hard ball model showing shells from the first to the fifth. Adapted from Kao et al. (2006).

Table 4.1 The Number of Atoms, Radius, and Relative Position for Each Shell in the Hard Ball Model

	First Shell	Second Shell	Third Shell	Fourth Shell	Fifth Shell
Number of atoms	6	6	6	12	6
Shell radius	$2r$	$3.46r$	$4r$	$5.29r$	$6r$
Mean radius of two close shells		$3.73r$		$5.65r$	
Deviation of shell from mean radius		−7.2%	+7.2%	−6.2%	+6.2%

Table 4.2 The Emergence of Close Shells and Atomic Size Difference for Equiatomic Alloys From NiAl to NiAlCuTiZrV Based on Hard Ball Model

Alloys	NiAl	NiAlCu	NiAlCuTi	NiAlCuTiZr	NiAlCuTiZrV
Atomic size fluctuation	± 6.7%	− 5.3%, +8.3%	± 7.8%	− 11%, +14%	− 10.4%, +14.7%
Deviations of second and third shells from mean	± 7.2%	± 7.2%	± 7.2%	± 7.2%	± 7.2%
Mergence of second and third shells	Partially merged	Partially merged	Partially merged	merged	merged
Deviations of fourth and fifth shells from mean	± 6.2%	± 6.2%	± 6.2%	± 6.2%	± 6.2%
Mergence of fourth and fifth shells	Merged	Merged	Merged	Merged	Merged

LAMMPS (Large-scale Atomic/Molecular Massively Parallel Simulator) package for MD simulations to find out the deposition and annealing processes of AlCoCrCuFeNi HEA thin films was adopted by Xie et al. (2013). Operational conditions in this simulation are magnetron sputtering of this HEA film on Si substrate. Cluster growth and thermal stability were studied during annealing in the range of 300–1500 K. The variation in root mean square displacement (RMSD) with increasing annealing temperatures is used for analysis of coalescence process. For example, at around 850 K a jump in RMSD value is observed indicating induced cluster coalescence. In computation, different methods were used for calculating interaction between different entities. For interactions among different elements in AlCrCoCuFeNi, embedded atom method force field was used. For Si–Si interactions and for Si–(AlCoCrCuFeNi) interactions, Tersoff potential and Lennard–Jones potential were used, respectively. Wang (2013b) generated Morse pair potentials for interactions among neighboring atoms using first principle calculations within DFT. This atomic model considers maximum entropy for the system. In this, using molecular statics simulation, optimized atomic configuration of AlCoCrCuFeNi alloy is found out. It is claimed that there is no short-range or long-range order in the alloys. The experimentally observed lattice structure of the alloy is only the result of average of the local disorder of atomic positions and composition in wide range.

4.2.4 MC Simulation

Similar to MD simulations, MC techniques are utilized for describing materials phenomena at the atomistic scale. The difference, however, is that while in MD simulations, the different microstates in the ensemble are traversed through physical forces describing the motion of atoms, in MC simulations these microstates of the system are derived through probabilistic arguments, which are based on the energetics of the given microstates. A useful application of the technique is in structure determination, in which multielement cluster configurations can be predicted using relatively simplistic MC computations.

Using this method, Wang (2013a) built atomic structure models of HEAs composed of four to eight elements in both BCC and FCC lattice according to the principle of maximum entropy (MaxEnt). Under the MaxEnt principle, every atom is striving for the maximum free space in the system. The driving force for this trend is called as the entropic force. The entropic force of atom i is pointed to the direction of $v_i(r)$ increase where $v_i(r)$ is the free space of atom i at position r. For example, consider a binary Fe–Cr alloy with a BCC lattice having $n \times n \times n$ unit cells in which the Fe atoms at some random sites are replaced by the solute atoms Cr. In the next step, this initial configuration is optimized according to the MaxEnt principle in the following way: The state of the maximum system entropy for this binary phase should be such that each of Cr atom approaches to its maximum free space. The main part of the optimization program for this is constituted by a big repeat loop. The position-replacing action is executed only when the distance of solute atom i, at the new position, to its nearest solute neighbor becomes longer than at its original one. The loop terminates when the MaxEnt condition is satisfied.

In the atomic structure models for HEAs of four to eight elements in both BCC and FCC lattices, by calculating the distance of an atom to its nearest same-element atom, Wang concludes that same elements as the first nearest neighbor can be avoided for those BCC solid solutions with five or more elements. Most of the same elements in quaternary and quinary BCC solid solutions are in the second nearest neighborhood (83.0% and 65.5%). This peak area moves to the third nearest neighborhood as the element number increases. On the other hand, the highest frequency of same-element existing as its first nearest neighbor for quaternary and quinary FCC solid solutions (74.3% and 47.3%). This peak moves to the second nearest neighbors for sexinary and septenary FCC solid solutions (64.9% and 50.4%), and the third one for octonary FCC solid solution (69.8%). Only a trace (0.2%) is left in the first nearest neighbor in the octonary phase. Wang further simulated FCC FeCoCrNi, FCC CoCrFeMnNi, BCC AlCoCrFeNi, and BCC AlCoCrCuFeNi alloys. In order to consider interatomic interaction in the simulation, Chen's lattice inversion pair-function potentials were used to obtain relaxed atomic structure models of the four alloys. In calculating lattice constant and lattice distortion, he found that BCC alloys generally have much greater lattice distortions than FCC ones. The reason is largely because the atomic radii of the component elements vary widely in the BCC phases.

This simulation demonstrates that the atomic structure with maximum entropy could be obtained for a multicomponent solid solution.

Lattice constant and distortion could also be calculated. Although the real alloys may deviate from this ideal state due to the diffusion resulting from the local unevenness of physical fields or the special affinity among some elements, the MaxEnt configuration can serve as the reference standard or the starting point for the studies of real materials.

4.2.5 Phase-Field Modeling

Over the years, one of the principle tools used for understanding the microstructure evolution during phase transformations is the phase-field modeling. The phase-field modeling is mainly used for solving interfacial problems and has mainly been applied to microstructure evolution during solidification (Boettinger et al., 2002). Phase-field models were first introduced by Fix et al. (1983) and Langer et al. (1986). Although, initially limited to solidification, it is also now widely used in areas such as solid-state diffusion, deformation behavior, heat treatment, recrystallization, grain coarsening, and so on. It has now become a handy tool for metallurgists and material scientists alike (Asta et al., 2009).

The phase-field method draws its elegance from the fact that it is able to describe microstructural evolution involving the motion of interfacial boundaries with concomitant coupling to heat, mass, and momentum transfer, without explicit tracking of the location of the phase boundaries. These problems are normally of the Stefan type and are usually quite complex to solve in the framework of classical finite element methods where the problems get more difficult with the increasing complexity of the boundaries. Traditionally, these problems were solved using the boundary-integral methods, which are numerically very efficient for 2D problems. However, these methods are unable to treat complex geometric pattern formation such as catastrophic phase termination, and additionally the equations of motion become quite involved in three dimensions. The phase-field simulations on the other hand retain their simplicity across dimensions and have the potential to be applied to problems of complex geometric pattern formation while retaining numerical simplicity and generality in the construction of the evolution equations of the phases.

Through the past two decades, the phase-field modeling techniques have been fine-tuned in order to derive quantitative information regarding measures of the microstructure such as lamellar spacings, dendritic arm spacings, and other morphological features such as dendritic tip radius as a function of different processing conditions. In particular, they have contributed in providing complete theories for the morphological evolution of dendrites and eutectics. These developments have naturally led them to being applied for the simulation of materials processing of real alloys involving multicomponent systems. In this endeavor, the coupling to thermodynamic databases is essential. While a number of examples of such coupling to databases exist, there are essentially two major pathways. One of them involves, the direct calculation of the thermodynamic driving force required in the evolution equation for the phase field, from the thermodynamic engine. This has been commercialized as the TQ interface (provided by Thermo-Calc), which hides all the complex thermodynamic calculations involving the computation of the driving force through a user-friendly interface. Since a single call to this TQ interface is generally time consuming, and given that the deviations in the driving force are normally small, the driving forces are extrapolated using the thermodynamic properties (liquidus slopes, Gibbs–Thomson coefficient, etc.) of the composition of the alloy. Timely checks are performed to ensure that the deviation of the extrapolation from the correct thermodynamic value is not significant. The commerical software (MICRESS, standing for Microstructure Evolution Simulation Softwares) combines the phase-field model to the thermodynamic database using such a coupling to the TQ interface. A second approach is the direct construction of simplistic free energy information developed out of the material properties (liquidus slopes, Gibbs–Thomson coefficient, partition coefficients) pertaining to the alloy composition one is modeling. The parametric determination of the coefficients of the constructed free energy is generally kept simple, such that they can be changed dynamically when there is significant change in the alloy composition.

There are no publications so far on the use of phase-field modeling in the area of HEAs. However, this unique modeling tool offers immense scope for understanding of microstructural evolution in these novel alloys. In the next few years, we can confidently expect more research so that the promise of multiscale modeling bridging the scales from electrons, through atoms, crystals, microstructure, and components, will be applied so that the dazzling but daunting possibilities of an enormous number of HEAs can be realized.

CHAPTER 5

Synthesis and Processing

5.1 INTRODUCTION

A variety of processing routes has been adopted for the synthesis of HEAs. HEAs have been synthesized in different forms like dense solid castings, powder metallurgy parts, and films. The processing routes can be broadly classified into three groups, namely, melting and casting route, powder metallurgy route, and deposition techniques.

Melting and casting techniques, with equilibrium and nonequilibrium cooling rates, have been used to produce HEAs in the shape of rods, bars, and ribbons. The most popular melt processing techniques are vacuum arc melting, vacuum induction melting, and melt spinning. Mechanical alloying (MA) followed by sintering has been the major solid-state processing route to produce sintered products. Sputtering, plasma nitriding, and cladding are the surface modification techniques used to produce thin films and thick layers of HEAs on various substrates.

This chapter gives a brief description of the different synthesis and processing routes adopted for HEAs. Processing routes are similar for both equiatomic and nonequiatomic HEAs.

5.2 MELTING AND CASTING ROUTE

The most widely adopted route for the synthesis of HEAs is the melting and casting route. Figure 5.1 gives an idea of the number of papers published on HEAs, grouped according to different synthesis routes. It is very clear from Figure 5.1 that the casting route (bulk) dominates the processing routes, with almost 75% of the papers published so far on HEAs being produced by this route.

A vast majority of HEAs that have been reported so far has been produced by vacuum arc melting and a few by vacuum induction melting. Arc melting has been the most popular technique for melting HEAs as the temperatures that can be achieved during arc melting are

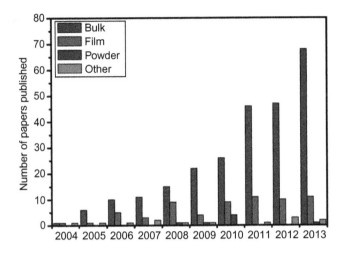

Figure 5.1 The number of papers published on HEAs that were produced by different processing routes.

high (close to about 3000°C), which is sufficient to melt most of the metals used for making HEAs. However, the disadvantage of this technique is the possibility of evaporation of certain low-boiling point elements during the alloy preparation thus making compositional control more difficult. In such cases, induction and resistance heating furnaces have been adopted for making the alloys.

One of the constraints faced in the melting and casting route is the heterogeneous microstructure developed due to various segregation mechanisms caused by the slow rate of solidification. The typical solidification microstructure of the HEAs produced by arc melting and casting is dendritic (DR) in nature with interdendritic (ID) segregation as shown in Figure 5.2 (Hemphill et al., 2012). Similar DR structure can also be seen in a number of other alloys prepared by induction melting and casting, as was observed in AlCoCrCuFeNi alloy (Singh et al., 2011b). The microstructure in this alloy showed DR and ID regions with a number of ordered precipitates (B2 and $L1_2$). Incidentally, when the same alloy was produced by a faster cooling route (splat cooling), it showed a BCC matrix with ordered (B2) precipitates. Cantor et al. (2004) also showed that melt spinning of a number of equiatomic quinary and sexinary alloys such as CoCrFeMnNi, CoCrFeMnNiNb, CoCrFeMnNiTi, CoCrFeMnNiV, CoCrFeMnNiCu, and CoCrFeMnNiGe leads to predominantly single phase FCC structure.

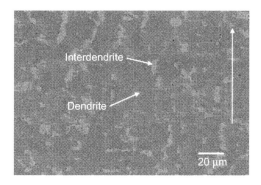

Figure 5.2 Microstructure of the as-cast Al$_{0.5}$CoCrCuFeNi alloy produced by arc melting and casting (Hemphill et al., 2012).

This demonstrates that faster cooling suppresses the precipitation of secondary phases leading to the formation of predominantly single-phase alloys. Among the melting and casting techniques, those that lead to faster solidification rates such as splat quenching, melt spinning, injection casting, suction casting, and drop casting have also shown similar microstructures with predominantly single-phase microstructures. This brings an important point to focus whether the single-phase structures obtained in some of the HEAs are kinetically favored or thermodynamically stabilized.

Cui et al. (2011a and 2011b) have directionally solidified AlCoCrFeNi and CoCrCuFeNi alloys, by vertical Bridgman technique. They observed that the microstructure of the alloy changes from planar to cellular and to DR on increasing the growth rate. Their results indicate that directional solidification can lead to finer DR structure with a decrease in the concentration difference between the DR and ID regions due to rapid growth rate and high-temperature gradients. Zhang et al. (2012b) observed, in the case of AlCoCrFeNi alloy, that the microstructure changes from DR to equiaxed when the alloy is prepared by Bridgman solidification in contrast to copper mold casting. Ma et al. (2013a) used extremely low withdrawal velocity of 5 μm/s to produce single crystals of FCC Al$_{0.3}$CoCrFeNi by Bridgman technique. In contrast, the equiatomic AlCoCrFeNi HEA under the same conditions yields columnar BCC grains, and single crystal could not be obtained in this case. The reason for this difference is yet to be understood.

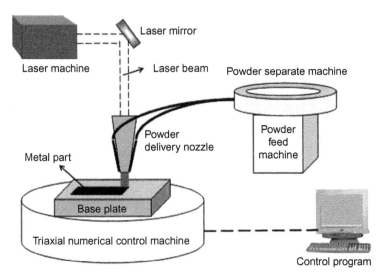

Figure 5.3 Schematic diagram depicting the LENS technique (Zhao et al., 2009).

Laser-engineered net shaping (LENS) in the technology of rapid prototyping can fabricate HEAs in bulk form directly by injecting metal powders into the area focused with high-powered laser beam. This technology was developed by Sandia National Laboratories for manufacturing solid metallic components from powder using a high-powered laser with a help of computer-aided design (CAD) model. Figure 5.3 shows a schematic of LENS technology. In this technique, the metal powder is fed through a deposition head placed coaxially to a focused laser beam. The X−Y table and the deposition head move with a number of degrees of freedom in order to generate the component with the required shape and size. An inert gas is used as a shield to prevent oxidation of the powder and the melt pool formed during the process. In developing HEAs, this technique has been used to produce gradient HEA rods layer by layer with changed compositions. For example, Al content can be varied from 0 to 3 segmentally in a grown $Al_xCoCrCuFeNi$ alloy rod (Welk et al., 2013). Similarly, other elements could be varied to produce segmentally gradient rods.

5.3 SOLID-STATE PROCESSING ROUTE

A small fraction of about 5% of the reports on HEAs so far deal with synthesis of HEAs by solid-state processing, which involves MA of the elemental blends followed by consolidation. MA is a process of

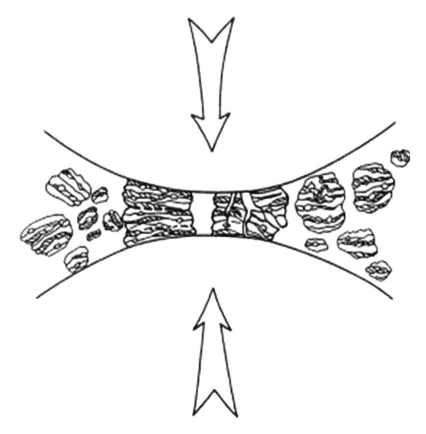

Figure 5.4 Fracture and welding phenomena during the collision of ball and powder particles during high-energy ball milling.

high-energy ball milling of elemental powder blends, which involves diffusion of species into each other in order to obtain a homogeneous alloy. This technique was first developed by Benjamin and his coworkers as a part of the program to produce oxide dispersion strengthened Ni base superalloys (Benjamin, 1970). In 1990, Fecht and his coworkers gave a first systematic report on the synthesis of nanocrystalline metals by high-energy ball milling (Fecht et al., 1990). Figure 5.4 shows schematically the ball to powder interaction during high-energy ball milling that involves continuous deformation, fracture, and welding of particles finally leading to the nanocrystallization or even amorphization. MA has been demonstrated over the past four decades as a viable processing route for the development of a variety of advanced materials such as nanomaterials, intermetallics, quasicrystals, amorphous materials, and nanocomposites (Murty and Ranganathan, 1998; Suryanarayana, 2001; Venugopal and Murty, 2011).

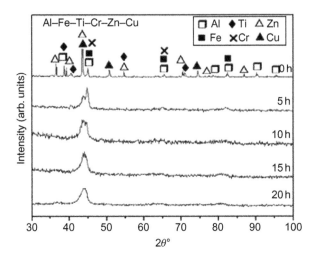

Figure 5.5 XRD patterns of equiatomic AlCrCuFeTiZn HEA obtained by MA (Varalakshmi et al., 2008).

The research group of Murty is the first to develop nanostructured HEAs using MA (Varalakshmi et al., 2008) and demonstrated high thermal stability and good mechanical properties of such alloys. Figure 5.5 (Varalakshmi et al., 2008) demonstrates the formation of nanocrystalline single-phase BCC after MA in a sexinary AlFeTiCrZnCu equiatomic elemental blend. One of the advantages of MA is its ability to produce excellent homogeneity in the alloy composition. Each of the nanoparticles obtained by MA is equiatomic in its composition, which has been confirmed by EDS and atom probe tomography.

These HEAs obtained by powder metallurgy route need to be sintered to achieve dense components. Conventional sintering of nanocrystalline alloy powders can lead to significant grain growth during the exposure of the alloy powders to high temperatures for long periods. In order to avoid this, nanocrystalline alloys obtained by MA are usually consolidated by spark plasma sintering (SPS). SPS involves application of high amperage pulsed current (up to 5000 A) through the sample kept usually in a graphite die, while simultaneously applying pressures to the tune of about 100 MPa as shown in Figure 5.6. The pulsed currents lead to the formation of spark plasma at the particle–particle interface in short periods causing almost instantaneous heating of the powder particles. This leads to sintering being completed within a short period of a few minutes in contrast to conventional sintering which takes a few hours to reach

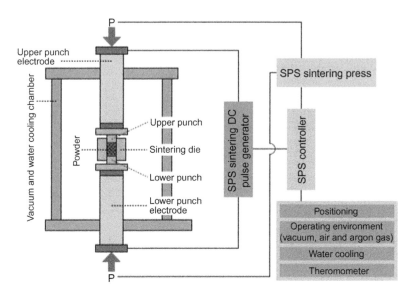

Figure 5.6 Schematic representation of operation of SPS. Courtesy: Fuji Electronic Industrial Co. Ltd., formerly known as SPS Syntex Inc., Japan.

similar sintering densities. Thus, the time available for grain growth during SPS is extremely small, which helps in retaining the nanostructures in the mechanically alloyed powder compacts. Retention of nanocrystallinity in CoCrFeNi alloy was observed even after annealing at 700°C up to 25 days subsequent to consolidation by SPS at 900°C.

5.4 HEA AND HEA-BASED COATINGS

Figure 5.1 shows that almost 20% of the papers on HEAs reported so far have been obtained in thin film/coating form by various techniques involving vapor and liquid.

5.4.1 HEA and HEA-Based Coatings from Vapor State

Among the vapor-based surface modifications, two techniques have been quite popular, namely, magnetron sputtering and plasma nitriding. The attempts by various investigators were to produce thin films or layers of HEA on the surfaces of substrates such as mild steels, Al alloys, and HEAs in order to improve corrosion resistance, oxidation resistance, and wear resistance.

Sputter deposition is a standard technique of depositing thin film onto a substrate by sputtering away atoms from a target under the

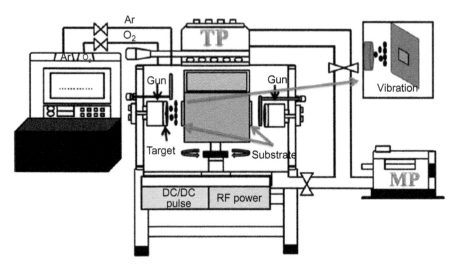

Figure 5.7 Schematic diagram showing the principle of DC and RF sputtering (Chen et al., 2013a).

bombardment of charged gas ions. DC sputtering shown in Figure 5.7 (Chen et al., 2013a) is the simplest of sputtering techniques wherein a DC bias is applied between the target and the substrate to aid the deposition. The deposition rates can be controlled by controlling power, the bias voltage, and the argon pressure. Radio frequency (RF) sputtering shown in Figure 5.7 (Chen et al., 2013a) is used for sputter deposition of insulating materials. In DC sputtering, if one attempts to sputter deposit an insulating film, a very high voltage to the order of 10^{12} V is required. This can be avoided in RF sputter deposition. In case of RF sputter deposition, the plasma can be maintained at a lower argon pressure than in DC sputter deposition, and hence fewer gas collisions leading to more lines of sight deposition.

In magnetron sputtering, electric and magnetic fields are used to increase the electron path length, thus leading to higher sputter deposition rates at lower argon pressures. The basic principle of magnetron sputtering is demonstrated in Figure 5.8. Magnetron sputter deposition uses both DC and RF for sputtering. Magnetron sputtering has been the most widely used coating technique for the HEAs (Tsai et al., 2013a).

Chang et al. (2008) have shown that DC magnetron sputtering of AlCrMoSiTi HEA in pure argon leads to amorphous thin film (Figure 5.9A). In the presence of nitrogen, the film turns out to be a simple NaCl-type FCC nitride (Figure 5.9B) mainly due to high-entropy effect. No phase separation of nitrides into TiN, AlN, Si_3N_4 is observed

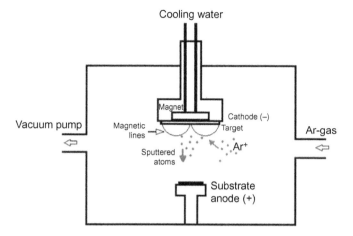

Figure 5.8 Schematic diagram showing the principle of magnetron sputtering (http://commons.wikimedia.org/wiki/File:Magnetronsputteren.png).

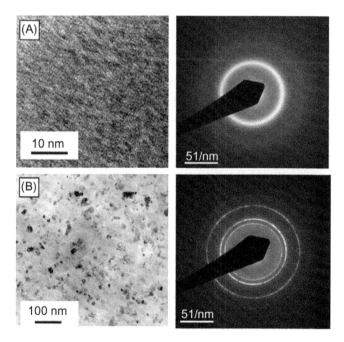

Figure 5.9 DC magnetron sputtered AlCrMoSiTi films in (A) argon and (B) nitrogen atmosphere (Chang et al., 2008).

in this system. An octonary AlMoNbSiTaTiVZr alloy has been successfully deposited as an amorphous film by Tsai et al. (2008b) which has a high performance as a diffusion barrier between Cu and Si in IC interconnects. Huang and Yeh (2010) have been able to develop superhard nitride coatings of AlCrNbSiTiV with hardness more than 40 GPa. Interestingly, it was found that the hardness of these coatings does not decrease with temperature even up to 1000°C. Grain coarsening was also observed to be minimal in these coatings up to 1000°C.

Similarly, sputtering (both RF and DC magnetron sputtering) of AlCrSiTiV alloy nitrides on mild steel substrate has shown a hardness of about 30 GPa and the grain size and hardness of these coatings were found to be quite stable even at 1173 K for 5 h (Lin et al., 2007). Similar results were observed by Chang et al. (2008) in case of AlCrMoSiTi nitrides. Braic et al. (2012) have recently developed HfNbTaTiZr nitride and carbide coating on Ti6Al4V alloy by DC magnetron sputtering for biomedical applications. They also observed that these coatings not only have excellent wear resistance but also have good biocompatibility in simulated body fluids.

Plasma nitriding is not as widely used as magnetron sputtering for making surface hardened layer for protection. Very few studies have been reported so far on this technique. However, this technique has been reported to produce thicker layer (50−100 μm) than magnetron sputtering (<1 μm). Plasma nitriding of $Al_{0.3}CrFe_{1.5}MnNi_{0.5}$ alloy has led to the formation of nitrided surface layer (Figure 5.10). The nitride layer has been analyzed as a mixture of various nitrides (AlN, CrN, and $(Mn,Fe)_4N$) and having a peak surface hardness around 1300 HV. By pin-on-disk adhesion wear test with an SKH-51 steel disc, the nitrided samples of HEAs with different prior processing have higher wear resistance than the unnitrided ones by 49 to 80 times and also than nitrided samples of conventional steels by 22 to 55 times (Tang et al., 2012). Feng et al. (2013) have recently combined magnetron sputtering with plasma-based ion implantation to produce NbTaTiWZr HEA-based nitrides with a better control on the thickness. They have also observed that this combination helps in improving the adhesion of the coating with the substrate.

5.4.2 HEA and HEA-Based Coatings from Liquid State

Various cladding techniques such as tungsten inert gas (TIG), also known as gas tungsten arc welding (GTAW), and laser cladding involve melting and casting of the coating material onto a substrate. The most common substrate for these cladding techniques has been mild steel. Chen et al. (2008) produced equiatomic AlCoCrMoNi alloy

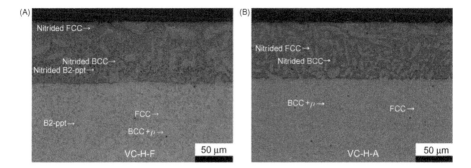

Figure 5.10 Optical micrographs of $Al_{0.3}CrFe_{1.5}MnNi_{0.5}$ alloy plasma nitrided at 978 K for 45 h: (A) the vacuum cast alloy homogenized at 1373 K for 2 h followed by furnace cooling and (B) the same alloy air cooled after the homogenization before nitriding treatment (Tang et al., 2012).

coating on low-carbon steel by TIG cladding. In this technique, the elemental powder blend of chosen alloy is used as filler material. During the process of TIG cladding, the filler material melts and picks up Fe from the substrate, and forms a cladded coating containing Fe in addition to the original filler composition. Hsieh et al. (2009) produced AlCrFeMnNi HEA coating by TIG welding process. In a similar way, Chen et al. (2009b) deposited AlCoCrFeMoNiSi HEA on low-carbon steel by GTAW. In both the above cases, the wear resistance of the cladded HEA was significantly higher than that of the substrate.

Huang et al. (2011) used laser cladding to produce AlCrSiTiV coating on Ti−6Al−4V substrate and reported that the coating resulted in an improvement in the oxidation resistance of the alloy at 800°C. In addition, the coating also showed improved wear resistance (Huang et al., 2012) due to the presence of hard silicides $(Ti,V)_5Si_3$ in the HEA coating.

5.5 COMBINATORIAL MATERIALS SYNTHESIS

Combinatorial chemistry uses chemical synthesis methods that make it possible to prepare a large number (up to even millions) of compositions in a single process. Combinatorial chemistry also includes strategies that allow identification of useful components of the libraries for such large-scale synthesis. Over the last two decades, combinatorial chemistry has altered the drug development process to discover new drugs (Pandeya and Thakkar, 2005). By this encouragement, materials scientists can also apply this methodology to accelerate the discovery of new compounds for high-T_c superconductors, luminescent materials,

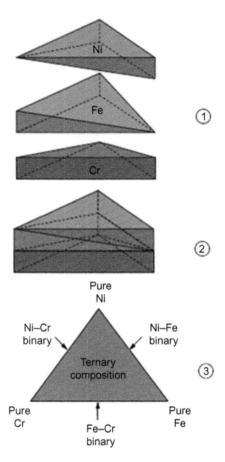

Figure 5.11 Schematic diagram showing the development of alloy library coupon using combinatorial materials science. Courtesy: Pharr, US Department of Energy, CPS#1778, 2006.

catalysts, and polymers (Xiang et al., 1995). They used thin-film technology to deposit substances sequentially in different amounts layer by layer onto a gridded substrate and then to mix the elements and create a stable compound by heating. The physical properties of interest are then measured on each composition to find out the outstanding composition. Basically under little guidance to predict new materials, this is a very efficient method to discover new materials in contrast to the conventional one-composition-at-a-time approach, which is time consuming.

For the development of multicomponent alloys by this method, the concept involves development of techniques that can fabricate large number of alloy specimen with continuous distribution of binary and ternary compositions across the surface, called the "alloy library." This technique saves the time, energy, and expense in alloy design and can help the development of new HEAs with improved properties. Figure 5.11 shows a schematic of the development of alloy library coupon using combinatorial materials science. In Figure 5.11, three controlled geometry thin films are deposited and annealed to develop one coupon with continuous distribution of elements. This high-throughput synthetic route holds great promise for further development of HEAs.

CHAPTER 6

High-Entropy Alloy Solid Solutions

6.1 INTRODUCTION

HEAs tend to form disordered and partially ordered solid solution phases having FCC, BCC, or HCP structures, instead of a mixture of many intermetallics. It is attributed to the high configurational entropy possessed by these alloys. For equiatomic alloys, the configurational entropy is given as $\Delta S_{\text{conf}} = R \ln(n)$ per mole, where n is the number of elements in equiatomic alloy and R is the universal gas constant. The high configurational entropy reduces the free energy of solid solutions in accordance with the relation $\Delta G_{\text{mix}} = \Delta H_{\text{mix}} - T\Delta S_{\text{conf}}$ and thus stabilizes the solid solutions, especially at high temperatures. This was first reported by Yeh and his coworkers, who showed that as-cast CoCrCuFeNi alloy has a single-phase FCC structure and AlCoCrCuFeNi alloy has a two-phase mixture of FCC and BCC structures. The BCC phase in the latter alloy spinodally decomposes into a modulated structure composed of ordered BCC (B2) and disordered BCC (A2) phases during cooling from the liquid state. They also emphasized that the phases at higher temperature are often solid solutions which might undergo phase transformation, such as spinodal decomposition (SD), ordering, or precipitation during cooling, because the importance of high mixing entropy in stabilizing solid solutions is reduced with decreasing temperature. In addition, sluggish diffusion will reduce the substitutional diffusion and lower the nucleation and growth rate and thus rate of phase transformation at lower temperatures (Yeh et al., 2004b). All these have been substantiated by several other works in later years. This chapter deals with solid solution phase formation, microstructural features, and thermal stability of these phases in equiatomic and nonequiatomic HEAs.

6.2 SOLID SOLUTION FORMATION IN EQUIATOMIC HEAs

Because materials with second-phase or multi-phase strengthening often have practical applications especially for those requiring high strength, high hardness, high-temperature softening resistance, and creep resistance, it is of considerable interest to identify the HEAs which can form a stable single phase and study their properties. A few conventional examples of single-phase alloys are α-brass, stainless steels, cupronickel, and nichrome. They have moderate strength and high ductility. A basic study of single phase HEAs leads to a better understanding of the mechanisms and properties of alloys containing more phases. The formation of a single-phase random solid solution is favored in the alloys where the majority of constituent binary pairs have a large mutual solid solubility and a small enthalpy of mixing. HEAs obtained through nonequilibrium processing routes like splat quenching, magnetron sputtering, and laser cladding usually exhibit single-phase structure because of lack of time available for the formation of second phases or even intermetallic phases. MA also promotes the formation of single-phase structure as it extends the mutual solid solubility of elements. Most of the HEAs contain either BCC, FCC, or a mixture of these two phases.

6.2.1 Solid Solutions with BCC Structure

BCC HEAs are important in high-strength applications as they usually show better mechanical properties like higher yield strength than FCC HEAs. Formation of BCC structure is favored when most of the binary pairs present in the alloy crystallize in BCC lattice. For example, AlCoCrFeNi alloy has been studied widely by a variety of processing routes such as arc melting and casting (Manzoni et al., 2013b; Wang et al., 2008), suction casting (Qiao et al., 2011), Bridgman solidification (Zhang et al., 2012b), and electro-spark deposition (Li et al., 2013c). Interestingly, this alloy shows a BCC/B2 structure irrespective of the processing route adopted, although only Cr and Fe have a BCC structure out of all the constituent elements. Zhang et al. (2012b) have shown that AlCoCrFeNi alloys processed by casting and Bridgman techniques, while having different solidification rates, show a single phase BCC structure. Interestingly, nanoparticles of B2 have been observed in the grains of BCC (A2) phase in this alloy made by Bridgman solidification.

Phase formation in HEAs appears to be governed more by the binary constituent pairs which evolve first rather than the individual elements themselves. This can be explained by considering phase evolution in AlCoCrFeNi alloy in greater detail. Qiao et al. (2011) reported that the framework of the crystal structure is Cr because of the closeness of lattice parameter of Cr (0.28847 nm) and that calculated for

AlCoCrFeNi alloy (0.289675 nm), and the fact that except Cr and Fe, none of the other constituent elements have BCC structure. However, it can be argued that the parent crystal structure is AlNi (B2 structure, which is an ordered structure based on BCC) rather than Cr. AlNi also has a lattice parameter of 0.28810 nm, which is also close to that of AlCoCrFeNi. The presence of superlattice peaks similar to that of AlNi in XRD patterns of AlCoCrFeNi also substantiates that the parent crystal structure is AlNi in which the other elements dissolve. The order parameter for the (Al,Ni)-rich phase will not be unity, as the crystal structure is disordered by substitution of other elements into the lattice. Further, considering the phase diagram of Cr (which is the first to solidify as it has the highest melting point) and Al (which has the highest diffusivity at solidification temperature as it has the lowest melting point) (ASM Handbook, 1992), Cr remains segregated from the liquid mixture up to $1350°C$ at the equiatomic composition. On the other hand, Al–Ni phase diagram indicates that at equiatomic composition solid AlNi begins to form at $1638°C$ itself. AlNi has the largest negative enthalpy of formation among all binary pairs in AlCoCrFeNi alloy, as shown in Table 2.6 (de Boer et al., 1988), remains stable over a wide composition field down to room temperature, and is able to dissolve other constituent elements. Thus, it is also an intermediate solid solution. In fact, Wang et al. (2014b) have investigated the structural evolution of $Al_xCoCrFeNi$ ($x = 0-1.8$) alloys in detail and concluded that AlCoCrFeNi alloys have a spinodal microstructure below 873 K, in which (Cr,Fe)-rich disordered (A2) solid solution and (Al,Ni)-rich (B2) solid solution form a modulated structure from high-temperature (Al,Ni)-rich (B2) solid solution. Zhang et al. (2012b) have prepared this alloy by Bridgman solidification and found fine B2 precipitates in BCC (A2) grains formed by SD (Figure 6.1). Higher withdrawal velocity under the same temperature gradient ~ 70 K/mm could lead to a finer modulated structure.

AlCoCrCuFeNi alloy has been produced by a number of processing routes. When the alloy is prepared by nonequilibrium techniques such as splat quenching (Singh et al., 2011b), DC magnetron sputtering (Dolique et al., 2009, 2010), and MA (Tariq et al., 2013; Zhang et al., 2009b,c), it shows only a single-phase BCC. When the alloy is prepared by more or less equilibrium routes such as arc melting and casting (Hsu et al., 2007; Tung et al., 2007; Yeh et al., 2007b; Wen et al., 2009), induction melting and casting (Kuznetsov et al., 2012, 2013; Singh et al., 2011a), and suction casting (Zhuang et al., 2012), Cu

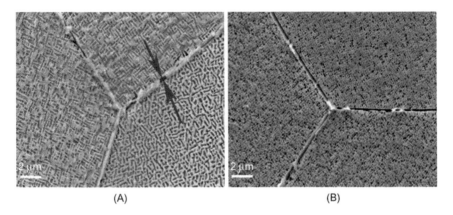

(A)   (B)

Figure 6.1 Microstructures of AlCoCrFeNi HEA prepared by Bridgman solidification with withdrawal velocities of (A) 200 and (B) 600 µm/s (Zhang et al., 2012b).

segregation is observed in ID region due to its positive enthalpy of mixing with other constituent elements as indicated in Table 2.6 (de Boer et al., 1988).

Therefore, the parent crystal structure observed in alloys containing significant amounts of aluminum and nickel is that of AlNi because it has the highest ordering energy among the binary pairs, which makes it form readily. In addition, other elements dissolve into the lattice subsequently due to their chemical compatibility and mixing entropy effect. The degree of ordering of the multicomponent alloy depends on the number and type of other elements present. This may be applicable to other HEAs as well, that is, a binary solid solution which forms most readily can evolve first and other elements dissolve into it subsequently.

Another interesting case to note is of AlCoCuNiTiZn which has been reported to form a single-phase BCC structure, although none of the constituent elements has a BCC structure at room temperature. Varalakshmi et al. (2010b,c) reported that this alloy can be formed by MA. Figure 6.2 shows the XRD pattern of AlCoCuNiTiZn with milling time which shows a single-phase BCC structure. Transmission electron microscopy (TEM) studies confirmed the nanocrystalline nature of as-milled AlCoCuNiTiZn alloy powder. The crystallite size was reported to be less than 10 nm after 20 h of milling. It is also important to note that the nanocrystallites of HEA obtained by MA have excellent homogeneity in their composition, which is demonstrated by the EDS spectrum (Figure 6.3) from one of the nanoparticles obtained in

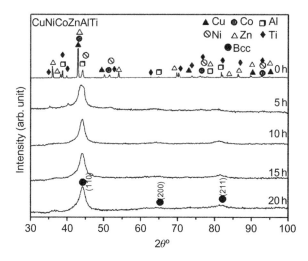

Figure 6.2 XRD patterns of AlCoCuNiTiZn alloy with different milling time (Varalakshmi et al., 2010c).

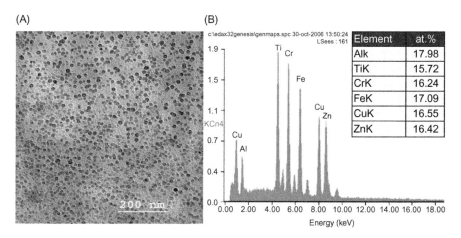

Figure 6.3 (A) Bright-field image of mechanically alloyed AlCrCuFeTiZn showing its nanocrystalline nature and (B) the EDS spectrum from one of the particles (Varalakshmi et al., 2008).

another AlCrCuFeTiZn HEA with BCC solid solution prepared by MA (Varalakshmi et al., 2008).

HEAs of refractory elements have been synthesized and are shown to have single-phase BCC structure in the as-cast state. Senkov et al. (2010 and 2011b) have reported that MoNbTaW and MoNbTaVW alloys exhibit BCC structure without formation of any complex phases or intermetallics. The lattice parameters reported for these alloys are

0.32134 and 0.31238 nm, respectively. Addition of vanadium reduces the lattice parameter of the quinary alloy as its lattice constant is the least (0.3039 nm) of all elements and also less than that of the quaternary alloy. This effect is similar when compared to the conventional alloys where incorporation of larger radii element causes lattice expansion and substitution of small radii element decreases the lattice constants. The formation of single phase is attributed to the conformance of individual elements to Hume-Rothery rules. Mo, Nb, Ta, V, and W have similar atomic radii, valency, and BCC crystal structure. Feng et al. (2012) have processed thin films of NbTaTiW which exhibit single-phase BCC structure. Although Ti has a HCP structure at room temperature, its BCC structure above 880°C, the complete solid solubility of Nb−Ta, Nb−W, and Ta−W and large solubility of Ti in Nb, W, and Ta lead to the formation of BCC crystal structure in this alloy.

From the above, it can be summarized that individual elements promoting BCC phase formation are Al, Cr, Fe, Ti, Mo, Nb, Ta, V, and W. Al is unique because it has a FCC structure. It has been shown that Al stabilizes FCC structure when its concentration is less than 11 at.% and promotes BCC phase formation when present in a greater amount (Wang et al., 2014b). This is because, although Al has a FCC structure, many of its binary compounds (AlNi, AlFe, AlCo, AlTi) crystallize in BCC lattice due to the formation of d−p hybrid orbital. Cr, Fe, Mo, Nb, Ta, V, and W have BCC structure at room temperature and hence tend to stabilize BCC structure in HEAs. Ti is HCP at room temperature but is BCC at higher temperatures and its large solubility in many BCC stabilizing elements like Nb, Ta, W, Al promotes formation of BCC HEAs. Table A1.1 in Appendix 1 gives a listing of various HEAs in which a single-phase BCC has been observed.

It should be mentioned that phase type prediction is done in a qualitative manner from constituent-element's features and/or unlike-atomic pair's features, which is similar to Hume-Rothery rules. While it is comprehensive, it is not possible to make an accurate prediction. This is simply because HEAs involve multielements and many unlike atomic pairs. As a result, various quantitative ways using parameters such as mixing entropy, mixing enthalpy, atomic size difference, and valence electron concentration are proposed for the prediction. Besides, various computational material science techniques including molecular dynamics simulation, *ab initio* calculation, and phase diagram calculation are proposed for pursuing direct outcomes. All these have been presented and discussed in Chapter 4. They are in fact complementary to each other for better understanding and prediction.

6.2.2 Solid Solutions with FCC Structure

FCC HEAs, owing to their closed-packed structure, are expected to show slower diffusion kinetics at elevated temperatures than BCC HEAs and hence are a better suited for high-temperature applications.

As discussed earlier, the formation of FCC phase is promoted when most of the binary constituents crystallize in FCC structure. For example, AlCoCrFeNi alloy shows an A2 + B2 structure. But, when Al is replaced by Cu to form CoCrCuFeNi alloy, FCC phase is formed (Cui et al., 2011b; Li et al., 2009). The binary constituents in CoCrCuFeNi alloy, namely, CoNi, CoFe, CuNi, CuCo, and FeNi, all have FCC structure and hence stabilize this phase. Similar behavior can be seen when Al is replaced by Mn to form CoCrFeMnNi alloy which exhibits single-phase FCC structure (Ye et al., 2012).

Varalakshmi et al. (2010b) studied the effect of elemental addition in HEAs. They synthesized CuNi, CuNiCo, CuNiCoZn, CuNiCoZnAl, and CuNiCoZnAlTi alloys by MA. Up to quinary alloys, a single-phase FCC structure is observed. CuNi is a well-known isomorphous system which crystallizes in the FCC lattice. Co is HCP at room temperature and it transforms to FCC above 450°C and is easily accommodated in the parent CuNi structure (FCC). Zn has a HCP structure, yet CoCuNiZn has a FCC structure which can be attributed to the presence of other three FCC elements (Co, Cu, and Ni). A particular case to notice is that of AlCoCuNiZn which crystallizes in a FCC lattice, even though Al content is 20 at.%. This indicates that formation of a CuNi isomorphous structure takes place readily before AlNi formation, and thus parent crystal structure is FCC, instead of B2, which is further stabilized by the presence of Co. In another similar interesting situation, Dolique et al. (2009, 2010) have observed the formation of a FCC phase in the DC magnetron sputtered AlCoCrCuFeNi alloy, when the deposited film has an Al content less than 15 at.% Al.

Praveen et al. (2012) have shown that CoFeNi and CoCuFeNi, formed by MA, show a single-phase FCC structure. It has been discussed that the parent structure in these alloys is that of Ni in which other elements dissolve. Complete solubility of Co−Ni and Cu−Ni and sufficient solubility of Fe in Ni stabilize the FCC structure in these alloys. Hence, whenever these elements are present together, we are more likely to get a FCC HEA except for a few cases where Al is present in larger amount.

Equiatomic alloys such as CoCrFeNi (Praveen et al., 2012) and CoCrFeMnNi (Liu et al., 2013; Otto et al., 2013a; Zhu et al., 2013; Tsai et al., 2013b) are important FCC alloys which predominantly exhibit a single-phase FCC structure. Here all the constituent elements are similar in sizes, valencies, and electronegativity. However, as discussed previously, uniqueness of HEAs lies in the fact that even elements of different crystal structure might combine to form a single-phase solid solution. For example, in CoCrFeNi and CoCrFeMnNi alloys mentioned above, not all the elements have FCC structure at room temperature, yet the alloys exhibit a single-phase structure. This non-conformance to Hume-Rothery rules in multicomponent system is, as discussed above for AlCoCrFeNi, because the phase evolution in HEAs is largely dominated by the binary constituent with highest driving force of formation. High-entropy effect enhancing the mixing of elements also has an important contribution.

Bhattacharjee et al. (2014) studied the texture development in CoCrFeMnNi alloy on deformation and annealing. They reported that the alloy exhibited a single phase FCC structure and developed submicron cell structure with a strong brass type texture on heavy cold rolling to 90% reduction. Annealing the alloy at 650°C showed ultrafine recrystallized grain structure with an average grain size of about 1 μm. In the recrystallized grain structure, a large fraction of annealing twins were formed. Remarkable resistance to grain growth up to 800°C and very large fraction of boundaries at misorientation angle about 60° (corresponding to the 60° <111> twinning relationship) have also been observed as shown in Figure 6.4. This study suggests that FCC HEAs generally have low-stacking fault energy and tend to have profuse twins and fine stable grain structure after rolling and annealing.

Table A1.2 in Appendix 1 lists the HEAs in which a single FCC phase has been observed.

6.2.3 Mixture of FCC and BCC Solid Solution Phases

As pointed out previously, nonequilibrium processing route is more likely to give rise to single phase. Most HEAs that show the presence of more than two phases are processed through equilibrium processing

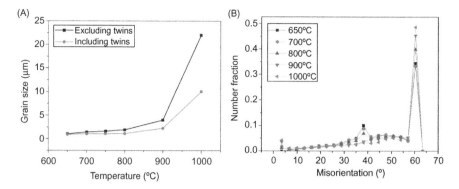

Figure 6.4 Variations of (A) grain size and (B) misorientation angle distribution with annealing temperature of CoCrFeMnNi HEA after 90% cold rolling (Bhattacharjee et al., 2014).

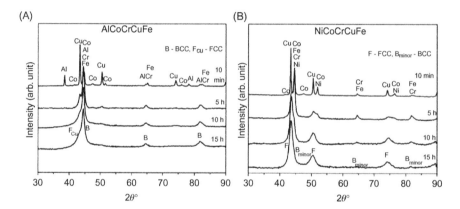

Figure 6.5 XRD patterns of (A) AlCoCrCuFe and (B) CoCrCuFeNi alloys with different milling time (Praveen et al., 2012).

routes like arc melting and furnace melting. The rate of solidification in these processes is slow and allows sufficient time for different phases to grow or for elemental segregation. For example, AlCoCrCuFeNi shows a single-phase structure when processed by MA or sputtering techniques but exhibits a mixed-phase structure (BCC + FCC) when synthesized through the arc melting route.

Praveen et al. (2012) discuss the phase formation behavior in AlCoCrCuFe and CoCrCuFeNi alloys processed by MA. The XRD patterns for these alloys are shown in Figure 6.5. AlCoCrCuFe crystallizes into a more open BCC structure which can accommodate an increased number of elements without much expansion, and hence no peak shift is observed after MA. On the other hand, the elemental peaks merge into Ni and peak shift of Ni has been observed in CoCrCuFeNi alloy (Figure 6.5B), which is explained on the basis of lattice expansion with increasing number of elements entering the basic Ni lattice during milling. Lattice expansion occurs as a consequence of more elements being accommodated in close-packed FCC structure. In addition, a minor BCC phase exists in CoCrCuFeNi alloy while a minor FCC phase exists in AlCoCrCuFe alloy. This is due to the difference in affinities between Al−Cu and Ni−Cu, which results in partial solubility and complete solubility of Cu during MA in AlCoCrCuFe and CoCrCuFeNi alloys, respectively. The presence of the minor FCC phase is attributed to Cu segregation in AlCoCrCuFe while Cr leads to the formation of the minor BCC phase in CoCrCuFeNi alloy. However, longer milling time is expected to further promote the mixing and chemical homogeneity.

Strong elemental segregations might occur when some elements have large positive mixing enthalpy in the interactions with other elements. Hsu et al. (2007) studied the effect of noble elements gold (Au) and silver (Ag) on phase formation in AlCoCrCuNi alloy. Ag, owing to its large positive enthalpy of mixing with other elements, leads to the formation of two layers and enhances Cu and Ag segregation in the liquid. The microstructure of these two layers along with EDX results shows that Ag segregates to Au layer and depleted in Ag layer. The Au layer enriched in Ag and Cu has a hypoeutectic structure of Ag-rich FCC phase and Cu-rich FCC phase whereas the silver layer enriched in Al, Co, Cr, and Ni has A2 + B2 structure. On the other hand, Au, due to its less positive enthalpy of mixing with other principal elements, combined well with the other elements and also reduced Cu segregation. Figure 6.6 shows DR microstructure of AlAuCoCrCuNi with EDX results showing good mixing of Au with other elements. In addition, XRD analysis shows two constituent phases of FCC + AuCu. This study also highlighted the importance of the mixing enthalpy between unlike atom pairs on the structural evolution and related properties of HEAs. Table A1.3 in Appendix 1 summarizes the compositions of HEAs in which a mixture of BCC and FCC phases has been observed.

6.2.4 Solid Solutions with HCP Structure

Very few HEAs are shown to crystallize in an HCP structure. Tsau and Chang (2013) reported the formation of HCP phase in quaternary

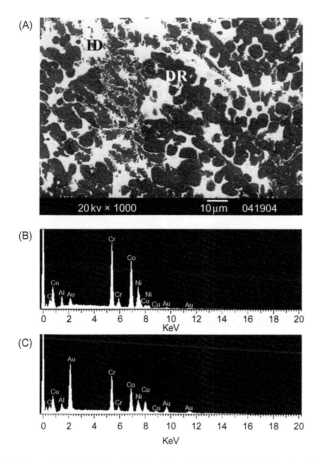

Figure 6.6 (A) SEM image of AlAuCoCrNi and EDX analysis of its (B) DR and (C) ID regions (Hsu et al., 2007).

TiCrZrNb alloy with a *c/a* ratio of 1.61; however, this phase is not present alone but is accompanied by a BCC phase with a lattice parameter of 0.298 nm. The HCP phase was present along the ID region while the BCC phase formed the matrix.

Huang et al. (2007) observed HCP phase formation in the oxide film of $AlCoCrCu_{0.5}FeNi$ HEA when the oxygen content is in the range of 10–50%. Tsau (2009) observed that the HCP phase gets stabilized when Ti is added to CoFeNi alloy. The as-cast CoFeNiTi alloy showed hard ordered HCP DR with a eutectic mixture in the ID region consisting of soft ordered FCC phase as matrix and hard ordered HCP phase as particles. The HCP phase is found to be stable on annealing up to 1000°C, while the FCC phase gets disordered

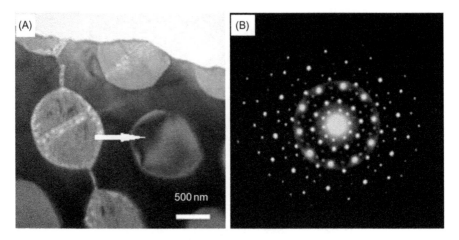

Figure 6.7 (A) TEM bright-field image and (B) SAD pattern from AlCuMgMnZn HEA showing the quasicrystalline phase (Li et al., 2011).

and softens on annealing for 2 h at 1000°C. Annealing of the alloy for 24 h at 1000°C has resulted in an ultimate compressive strength of 2.60 GPa and a plastic strain of 20%.

In another interesting study, Li et al. (2010b) showed the formation of an HCP phase along with Al−Mn type quasicrystalline phase in as-cast $Mg_x(AlCuMnZn)_{100-x}$ ($x = 20$, 33, 43, 45.6, and 50) alloys. It is important to note that the quasicrystalline phase in this alloy has been observed without the need for rapid solidification. In another report, Li et al. (2011) have provided an electron diffraction evidence for the quasicrystalline phase formation (Figure 6.7) along with an HCP phase in AlCuMgMnZn HEA.

Shun et al. (2010b) observed an ID HCP phase and DR FCC phase in $Al_{0.3}CoCrFeNiTi_{0.1}$ alloy. Shun et al. (2010a) have also seen similar (Ni,Ti)-rich HCP phase in the ID regions of $CoCrFeNiTi_{0.3}$ alloy. Tsai et al. (2010c) have reported an HCP phase in the CrTiVZrY HEA coatings obtained by magnetron sputtering. The nitride coatings obtained from this HEA have shown NaCl-type FCC structure. The HCP phase in the HEA coating was attributed to an HCP structure of three of the five elements (Ti, Zr, and Y) in the alloy. In an attempt to search for a single-phase HCP, Gao and Alman (2013) used CALPHAD approach to predict the HCP phase formation in CoOsReRu equiatomic alloy. However, they did not provide any experimental evidence for this prediction.

6.3 SOLID SOLUTION FORMATION IN NONEQUIATOMIC HEAs

The original concept of designing HEAs used the principle that configurational entropy, given by $\Delta S_{conf} = -R\Sigma X_i \ln X_i$ per mole, is maximum for $X_i = 1/n$, where X_i is the mole fraction of the ith element, R is the universal gas constant, and n is the number of elements in the alloy. Nonequiatomic alloys have a lower configurational entropy as their composition deviates from equiatomic one. This implies that tendency for solid solution formation would be the greatest for an equiatomic HEA. Yet, we see a large number of researchers are working on nonequiatomic HEAs. The reasons for this are threefold. Firstly, the difference between configurational entropy of equiatomic and nonequiatomic HEA is not very large. For example, configurational entropy (ΔS_{conf}) of six-component AlCoCrCuFeNi alloy comes out to be $1.93R$, while ΔS_{conf} for $Al_{0.2}CoCrCuFeNi$ alloy would be $1.78R$. Secondly, many nonequiatomic HEAs have been shown to have superior mechanical properties than the corresponding equiatomic alloys. Thirdly, it has been of interest to determine the effects of various elements on structural evolution and properties of HEAs. Yeh et al. (2004b) studied the effect of aluminum in as-cast $Al_xCoCrCuFeNi$ alloys and showed that aluminum when added in smaller quantities (<0.5), the alloy has a FCC structure but promotes BCC phase formation when its content is higher. This helps in designing HEAs with required microstructure and properties.

The main constituent phases in nonequiatomic alloys are not significantly different from that of the equiatomic alloys. It is also very probable to find single-phase HEA from nonequitaomic composition. For example, equiatomic AlCoCrCuFeNi HEA has a mixture of FCC and BCC besides Cu-rich ID phase whereas nonequiatomic $Al_{0.3}CoCrCu_{0.4}FeNi$ HEA has a single FCC phase and no Cu-rich ID phase. Similarly, $CoCrCu_{0.5}FeNi$ forms a single-phase FCC structure (Lin et al., 2010). $Al_xCoCrFeNi$ forms an ordered BCC structure even when x is higher than 1 (Li et al., 2009). According to Gibbs phase rule, for a sexinary alloy, there are six degrees of freedom in the composition range for the single-phase field. As a result, a large number of nonequiatomic compositions can give raise to a single phase. However, in some cases, these alloys might show typical DR microstructure with ID regions having different composition in comparison to the DR. This DR structure is related to the freezing range as seen in conventional as-cast alloys.

Obviously, nonequiatomic HEAs might also have two or more phases. AlCoCu$_x$NiZnTi exhibits a simple solid solution structure (FCC + BCC) for different copper contents (Varalakshmi et al., 2010b). In all these cases, the phases observed are similar to those observed for corresponding equiatomic HEAs. Ren et al. (2012) studied a series of nonequiatomic HEAs containing Cr, Cu, Fe, Mn, and Ni by varying two elements at a time. In all the alloys, they observed either FCC or FCC + BCC structures without the formation of any complex phases or compounds.

The influence of processing routes on nonequiatomic HEAs is similar to that of equiatomic HEAs. Nonequilibrium processing routes like MA and rapid solidification techniques are more likely to give raise to single-phase structures whereas equilibrium synthesis techniques like casting, in some cases, can lead to the formation of intermetallic phases. Al$_{0.5}$CoCrCuFeNi exhibits a single-phase FCC structure when prepared by magnetron sputtering (Chen et al., 2005a) whereas the formation of ordered L1$_2$ phase is observed when the same alloy is processed through arc melting route (Hemphill et al., 2012). Huang et al. (2004) showed that AlCrFeMo$_{0.5}$NiSiTi exhibits a mixed-phase structure containing one B2 phase and two FCC phases when processed through arc melting and casting, while it shows almost a single-phase BCC structure when prepared by plasma spraying.

6.3.1 Effect of Aluminum

Aluminum has been a major alloying element in most of the HEAs studied till date. It is believed to impart strength and good oxidation resistance. At lower Al concentrations, it stabilizes FCC phase since it has the same structure. When added in higher amounts, it tends to stabilize ordered BCC (B2) structure which can be attributed to the fact that it forms stable binary compounds with many of the common elements like AlNi, AlFe, AlCo, etc. The XRD studies of Al$_x$CoCrCuFeNi alloys prepared by arc melting and casting by Yeh et al. (2004b) showed this trend. The alloy with Al content between 0 and 0.5 has an FCC structure, but a BCC phase appears after $x = 0.5$. The amount of BCC phase increases after this, but the FCC phase starts decreasing after $x = 2.8$. At higher concentration, the FCC phase disappeared and a single-phase B2 structure was observed. Similar results have been reported by Chou et al. (2009) for Al$_x$CoCrFeNi alloys (Figure 6.8). The alloy shows FCC structure

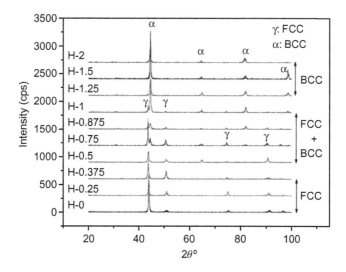

Figure 6.8 XRD patterns of $Al_xCoCrFeNi$ alloys for different x values (Chou et al., 2009).

with x up to 0.375. For alloys with $x = 0.5-1.0$, the alloy shows a FCC + BCC structure. When the Al content is higher than $x = 1.25$, the alloy has a BCC structure. Hsu et al. (2013b) have shown that addition of Al not only promotes the formation of B2 structure but also hampers the evolution of σ phase in $Al_xCoCrFeMo_{0.5}Ni$ alloy. This can be attributed to the fact that most of the elements have negative interaction energy with Al and binary constituents like AlNi can readily evolve which further can dissolve other σ phase forming elements like Co, Cr, and Fe.

The morphology of cast structure of HEAs also changes with the addition of aluminum. Figure 6.9 shows the different microstructures observed when Al content is varied from 0 to 2 for $Al_xCoCrFeNi$ alloys (Wang et al., 2012b). The cast structure is cellular for $x = 0-0.3$, columnar DR for $x = 0.4-0.6$, equiaxed non-DR for $x = 0.7-0.8$, equiaxed DR for $x = 0.9-1.5$, and nonequiaxed DR for $x = 1.8-2$.

In a recent report, Tang et al. (2013) discuss the mechanism of interaction between Al and transition metals in HEAs. Aluminum has three electrons in the outer shell, a small work function, and a low ionization energy. Consequently, it prefers to transfer electron to vacant d orbitals of transition metals like Fe, Co, Ni, Cr, and Mn to form strong covalent

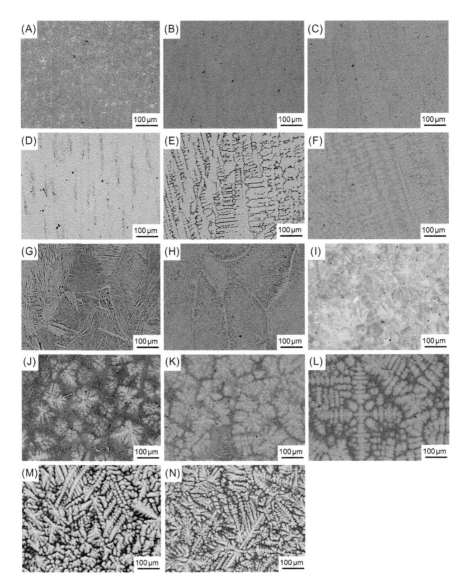

Figure 6.9 Optical micrographs of Al_xCoCrFeNi alloys for different Al content (Wang et al., 2012b).

bonds and intermetallic compounds. In addition, Al, being a larger atom, causes a lattice distortion in crystal, particularly in HEA consisting of transition metals. The lattice distortion can be better accommodated if the lattice has a more open structure than a close-packed one. Therefore, it is seen that BCC phase formation is favored at higher Al contents.

6.3.2 Effect of Transition Elements (Co, Cr, Cu, Mo, Ni, Ti, and V)

Transition elements form important constituents of HEAs. They stabilize the HEA phase similar to their own crystal structure. Co (above 450°C), Cu, and Ni are FCC elements and stabilize FCC phase in HEAs while Cr, Mo, V, and Ti (high-temperature BCC) favor the formation of BCC structure. Wang and Zhang (2008) showed that increasing the cobalt content in $AlCo_xCrFeNiTi_{0.5}$ alloy beyond $x = 1$ leads to the appearance of FCC phase which increases in its intensity with further increases in Co content.

Copper tends to segregate in ID region due to its positive enthalpy of mixing with many common elements. This is sometimes reflected in the XRD patterns of HEAs with the presence of small FCC peak. The amount of copper decides the mechanism of segregation and hence influences the microstructural features of the alloys. Mishra et al. (2012) studied the phase evolution in Co−Cu−Fe−Ni−Ti alloys with different Ti/Cu atomic ratios. At lower ratios (Ti/Cu = 9/11, 11/9, and 3/2), Cu-rich liquid segregates at ID region subsequently causing formation of eutectic mixture of Cu-rich phase and Laves phase of Ti_2Co type. At higher atomic ratios (Ti/Cu = 1/3, 3/7, and 3/5), two solid solution phases, one Co-rich and other Cu-rich phase, are observed. This occurs due to phase separation between Cu-rich and Co-rich liquid owing to their positive enthalpy of mixing. Formation of two solid solutions has also been observed in $Cu_xZn_yTi_{20}Fe_{20}Cr_{20}$ (Mridha et al., 2013) alloys synthesized by MA for different x/y ratios. One solid solution which is Cr-rich has BCC structure while the other solution is Cu-rich and has FCC structure.

Compared to Co, Ni is a stronger FCC stabilizer. In fact, Ni is the strongest FCC stabilizer among all FCC stabilizers. Addition of Ni causes appearance or stabilization of FCC phase in HEAs. XRD studies of $AlCoCrFeMo_{0.5}Ni_x$ alloys for different x values by Juan et al. (2013) revealed that FCC phase appears at higher nickel content, confirming that Ni stabilizes FCC structure.

Addition of molybdenum tends to stabilize the formation of BCC structure and/or appearance of σ phase. When added to $AlCrFeNiMo_x$ alloys, Mo dissolves preferentially in Fe−Cr BCC phase. Cr is an important constituent in HEAs, and as already discussed in the context of equiatomic alloys, Cr stabilizes BCC structure and promotes formation of σ phase particularly in presence of Fe, Co, and Ni. Chen et al. (2006a) discussed the effect of vanadium on FCC $Al_{0.5}CoCrCuFeNi$ alloy. V when present in higher concentration (>0.2) causes phases separation and leads to a BCC phase with modulated structure by SD, which envelops the FCC DR.

Ti crystallizes in an HCP structure at room temperature which transforms to a BCC structure at higher temperatures. Ti is often added to improve corrosion resistance of alloy and increase strength by solid solution strengthening or precipitation hardening. Ti is usually found to favor the formation of BCC structure.

6.4 MICROSTRUCTURE OF HEAs

HEAs processed through a casting route show typical cast microstructure consisting of DR and ID. DR region is often found to contain microstructural features like precipitates, nanostructured phases, and modulated structure arising from SD. Elements like Cu and Ag have been found to segregate in ID region of cast microstructure. ID regions are also shown to have two-phase eutectic structure. Tong et al. (2005b) have studied $Al_xCoCrCuFeNi$ HEAs ($x = 0-3.0$) and pointed out that the occurrence of precipitation is because of the decreased solubility of the FCC and BCC matrix phases with lowering temperature due to the diminishing effect of high mixing entropy. Figure 6.10 (Tong et al., 2005b) depicts phase formation sequence during cooling of $Al_xCoCrCuFeNi$ alloy system with different aluminum contents.

Tung et al. (2007) discussed microstructural features of various HEAs having different elements in nonequiatomic proportion. Alloys having lesser content of copper have narrow ID regions which is evident from the segregation tendency of the element. SD leading to modulated structures is observed in the alloys containing BCC phase (DR regions) while ID regions have mixed FCC and BCC structures. It can be readily inferred from these results that although configurational entropy for different alloys is same, their microstructures vary in terms of phase fractions and compositions which further reinforces the fact that other thermodynamic factors also play a role in phase evolution in nonequiatomic HEAs.

Singh et al. (2011b) have studied microstructure of AlCoCrCuFeNi alloy in detail (Figure 6.11). They have observed the DR microstructure in as-cast alloy (Figure 6.11A). It has been shown that DR region

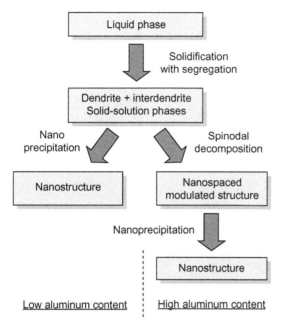

Figure 6.10 Depiction of phase formation sequence during cooling of Al$_x$CoCrCuFeNi alloy system with different aluminum contents (Tong et al., 2005b).

actually comprises of many minor phases like plate-like precipitates (Figure 6.11B), rhombohedral and spherical precipitates (Figure 6.11C), and weak superlattice reflections of L1$_2$ phase. The DR region was also shown to have Ni−Al, Cr−Fe, and Cu-rich plates as studied by 3D atom probe (Figure 6.12). On the other hand, splat-quenched AlCoCrCuFeNi alloy exhibited a polycrystalline microstructure with clear grains and grain boundaries. Figure 6.13 shows a high-angle annular dark-field (HAADF) image along with elemental EDS maps, obtained using image corrected TITAN, of Al$_{1.5}$CoCrCuFeNi alloy prepared from elemental powders with a LENS device (Welk et al., 2013). Ni, Al, and Co appear to be more concentrated in one of the plate-like structures, while Cr and Fe are concentrated in the other plate-like phase. The cylindrical- and elliptical-shaped phases appear to be very rich in Cu. Furthermore, the nanoscale precipitates are also rich in Cu. This fine structure again confirmed the SD. The first of these plate-like phases corresponds to the B2 phase whereas the other plate-like phase corresponds to the disordered BCC (A2) phase.

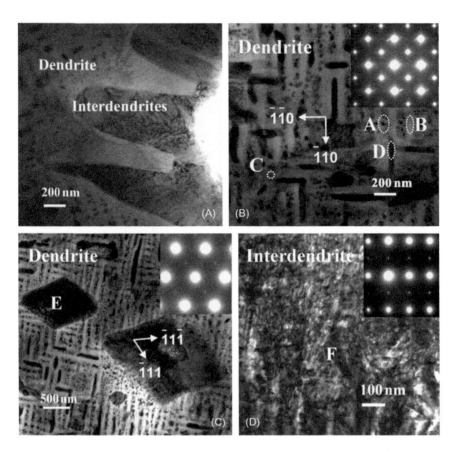

Figure 6.11 Bright-field TEM images showing (A) DR and ID regions, (B) DR showing plate-like precipitates and presence of ordered B2 structure, (C) presence of rhombohedral precipitates in DR and weak reflections of L1₂ phase, and (D) microstructure of ID region and weak superlattice reflections of L1₂ phase for as-cast AlCoCrCuFeNi alloy (Singh et al., 2011b).

In contrast to the above observation on cast alloys, atom probe studies on mechanically alloyed CoCrFeNi HEAs indicate very uniform distribution of alloying elements in the as-milled condition. However, segregation of certain elements in some alloys prepared by MA has been observed after hot consolidation. Elemental mapping using atom probe tomography in AlCoCrCuNiZn HEA prepared by MA followed by hot compaction at 600°C showed segregation of Cu to the grain boundaries. Retention of nanocrystalline grains of about 10 nm in size even after consolidation was also observed.

In addition to processing routes, microstructure of HEAs also depends on the alloying element because the phase equilibrium and kinetics in

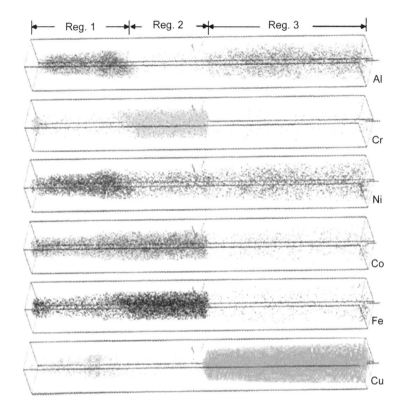

Figure 6.12 3D reconstruction of Al, Cr, Ni, Co, Fe, and Cu atom positions in AlCoCrCuFeNi alloys (Singh et al., 2011b).

solidification stage are also changed. For example, Ti addition to AlCoCuFeNi changes morphology from DR to eutectic cell type (Wang et al., 2012c) whereas V addition exhibits DR region with ellipsoidal particles instead of modulated plate-like structure. They have observed morphology to change from DR to columnar on Ti addition.

Zhang et al. (2013a) reported the formation of lath-like martensite containing high density of dislocations in laser-solidified FeCoNiCrCuTiMoAlSiB$_{0.5}$. Such martensite has been commonly observed in steels and shape memory alloys but not in HEA systems. The nucleation of martensite is mainly due to high cooling rates during laser-solidification process. In addition, boron atoms are believed to play a similar role as carbon does in steels. Boron atoms, being smaller in size, sit in octahedral voids and result in contraction and expansion along the shear modulus direction causing martensitic

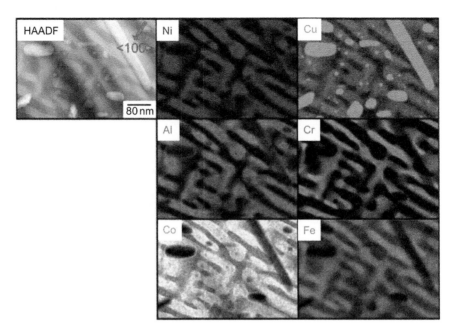

Figure 6.13 An HAADF image along with elemental EDS maps, obtained using image corrected TITAN, from Al₁.₅CoCrCuFeNi alloy (Welk et al., 2013).

transformation. Interestingly, the martensite transforms to an ordered B2 structure for which the driving force is probably the reduction in strain energy and enhancement of crystal symmetry above a critical limit of solute concentration.

Advanced characterization techniques like high-resolution transmission electron microscopy, atom probe tomography, electron back-scattered diffraction (EBSD), etc. prove to be very helpful in observing the fine microstructural details of HEAs. Figure 6.14 (Yang et al., 2011) demonstrates the use of EBSD in the study of the interface between Al and AlCoCrFeMnNi HEA.

Mechanically alloyed powders of nonequiatomic HEAs show similar structures as that of equiatomic HEAs. Hard agglomerates containing nanocrystallites, smooth fine particles, and uniform morphology are some of the characteristic features of the microstructure. Chen et al. (2009c) prepared AlCoCrCu₀.₅FeMoNiTi alloy by MA. Consistent with the mechanisms of MA, the 2 h-milled powders show lamellar structures which transform to uniform featureless microstructure after 36 h of milling. In a similar study by Sriharitha et al. (2013) on Al₅CoCrCuFeNi alloy

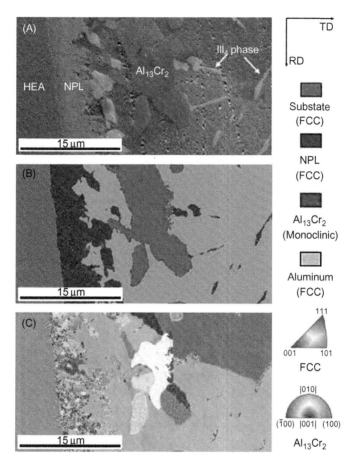

Figure 6.14 (A) SEM image of the Al−AlCoCrFeMnNi HEA interface, (B) phase map of the Al/AlCoCrFeMnNi HEA interface obtained from EBSD, and (C) the corresponding orientation maps (Yang et al., 2011).

prepared by MA, the microstructure showed an average particle size of around 0.5 microns with a crystallite size less than 50 nm.

6.5 ROLE OF SLUGGISH DIFFUSION IN PHASE EVOLUTION OF HEAs

One of the distinct features of HEAs is the slow rate of diffusion due to the presence of a large variety of elements. This sluggish diffusion in HEAs not only improves the high-temperature properties but also plays a significant role in their phase evolution and thermal stability.

Varalakshmi et al. (2008) reported the synthesis of nanocrystalline BCC HEAs having two to six components. Among the AlFe, AlFeTi,

AlFeTiCr, AlFeTiCrZn, and AlFeTiCrZnCu alloys made by MA of individual elements, it has been observed that phase formation is complete after longer milling time in quinary and sexinary alloys when compared to binary to quaternary alloys. It is attributed to sluggish diffusion kinetics with increasing number of elements. The slow rates of diffusion are also responsible for the formation of nanostructures in HEAs. Cantor et al. (2004) report that primary DR arm width (15 μm) seen for quinary CoCrFeMnNi alloy is larger than that (5 μm) observed for sexinary alloy CoCrCuFeMnNi alloy. In the same work, it is shown that when Nb and Ti are added to CoCrFeMnNi alloy, a single phase is still retained. Nb and Ti, both are not FCC elements and tend to stabilize BCC phase, yet they dissolve completely into single-phase FCC HEA. The slow diffusion rates in the alloy possibly prevents the formation of any second phase or complex precipitates as observed with Nb, Ti addition in conventional alloys.

The current understanding of mechanisms responsible for sluggish diffusion in HEAs is incomplete. The very first paper reporting diffusion measurements in HEAs appeared in 2013 (Tsai et al., 2013b). Pseudo binary approach using diffusion couple technique was adopted to measure interdiffusion coefficients in CoCrFeMnNi alloy. The diffusion coefficients obtained for different elements in HEAs are much lower when compared to that in conventional alloys. For example, Cr and Ni diffusion coefficients in CoCrFeMnNi alloy are calculated as 1.69×10^{-13} and 0.95×10^{-13} m^2/s, respectively, whereas the values of these coefficients in FCC-Fe are 4.19×10^{-13} and 2.66×10^{-13} m^2/s, respectively. The origin of slow diffusion rates is discussed in terms of varied interaction energies, also termed as Lattice Potential Energy (LPE) at different sites due to the presence of large number of elements in higher concentration. The larger LPE fluctuation in HEA leads to higher normalized activation energies and a lower diffusion rate, and thus the sluggish diffusion effect.

The sluggish diffusion can also be explained in terms of reduced activity of each element in the single-phase solid solution. Presence of different kinds of elements creates additional strain in the lattice, which further enhances the barrier for atom and vacancy migration. A detailed study of diffusion mechanisms by techniques like tracer diffusion, experimental determination of vacancy concentration by positron annihilation, etc. can provide better understanding of this field.

6.6 THERMAL STABILITY OF HEAs

HEAs have been explored for applications at high temperatures. Their performance as coating and structural materials at high temperatures depends on their creep behavior, stability of microstructure, and mechanical properties at elevated temperatures. Therefore, it is of critical importance to understand the thermal stability of phases and microstructure of HEAs.

Alloys processed through nonequilibrium processing routes like MA and sputtering show formation of metastable phases which may transform to new structures when subjected to thermal treatments. On the other hand, processing routes like casting result in the formation of equilibrium phases, consequently most of the cast alloys retain the solid solution phases on heat treatment. In some cases, however, a new second phase may evolve. Phase evolution upon heat treatment has been found to depend broadly on alloy composition, processing conditions, heat treatment process, and initial phases. Alloys having only a single solid solution phase in as-processed condition are more likely to be stable when subjected to heat treatment than alloys having mixed phases. If several phases are involved for strengthening and creep resistance purposes, the reinforced phases are also required to be thermally stable under operating condition as found in superalloys, for example, stable gamma prime in the gamma matrix is very important.

Varalakshmi et al. (2010c) have studied the effect of consolidation process for AlFeTiCrZnCu alloy. It has been shown that as-milled alloy contains a BCC and FCC phases, which transforms to two BCC phases depending upon consolidation process. Although the XRD patterns of HIPed (Hot Isostatic Pressed) and VHP (Vacuum Hot Pressed) are similar, the particles in HIPed samples are relatively smaller in size and spherical in shape due to application of uniform pressure from all sides.

The study of phase evolution in NiCoFe, NiCoCrFe, NiCoCuFe, NiCoCrCuFe, and AlCoCrCuFe after spark plasma sintering (SPS) at 900°C (Praveen et al., 2012) has demonstrated that Cu segregation (indicated by evolution of FCC peak) and formation of σ phase occur in AlCoCrCuFe and NiCoCrCuFe. In AlCoCrCuFe, the major BCC phase transforms to an ordered BCC phase. Formation of σ phase is observed in all Cr-containing alloys.

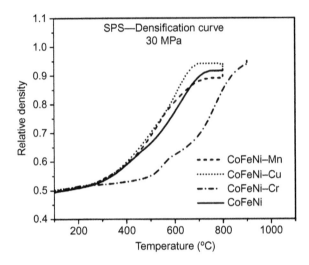

Figure 6.15 Densification curves for different mechanically alloyed alloys during SPS (Praveen et al., 2013b).

Praveen et al. (2013b) discuss the densification kinetics of CoCrFeNi alloy in detail. Figure 6.15 shows the densification curve obtained for different quaternary HEAs which indicates a delayed densification in CoCrFeNi alloy. The CoCrFeNi alloy shows a minor BCC phase in the XRD pattern. This phase transforms to a σ phase around 550°C, after which densification is also found to accelerate. BCC phase is a Cr-based phase and hence delays the densification of HEA because of slow diffusion rate of Cr owing to its high melting point. This alloy also demonstrated excellent stability for the nanocrystalline nature of the alloy and the crystallite size remained around 50 nm even after annealing for 25 days at 700°C, after SPS at 900°C.

One important difference that is observed in many HEAs when compared to conventional alloys is the nature of room temperature phase vis-a-vis the high-temperature phase. In conventional elements and alloys, a close-packed structure is more stable at room temperature and a more open structure is observed at higher temperatures. Also in conventional alloys an ordered phase is observed at lower temperature whereas disordered phase appears at higher temperature. However, for some HEAs a reverse trend is observed in both the cases (Praveen et al., 2012; Zhang et al., 2009b; Dolique et al., 2010). As discussed in previous section, crystal structure of HEAs is characterized by the presence of high strain due to severe lattice distortion.

This strain is higher for a close-packed structure. Therefore, at room temperature, atoms tend to crystallize in a more open structure so that additional strain can be compensated. At higher temperature, the extra strain appears to be compensated, making the close-packed structure more stable.

MA usually leads to the formation of metastable phases which often disappear or transform to other phases at higher temperatures. Metastable phases are observed due to inherent nonequilibrium nature of the process. Sriharitha et al. (2013) studied the influence of heat treatment after MA of $Al_xCoCrCuFeNi$ alloys. They observed three phases, a major BCC and two minor FCC, in $Al_{0.45}CoCrCuFeNi$ in the as-milled samples. After differential scanning calorimetry (DSC) at 1480°C, the BCC phase disappears and two FCC phases form the structure. Formation of a new FCC phase is observed when $Al_{2.5}CoCrCuFeNi$ alloy is subjected to DSC. $Al_5CoCrCuFeNi$ retains its single-phase B2 structure even after DSC which suggests high thermal stability for the alloy. Fu et al. (2013a) reported that $Al_{0.6}CoCrFeNiTi_{0.4}$ shows BCC and FCC phases when processed through MA, and the phases transform to new BCC and FCC structures when subjected to SPS.

Alloys produced by casting route usually have equilibrium structures and hence are less likely to transform to new ones when subjected to higher temperatures. However, the presence of certain alloying elements can lead to precipitation and evolution of phases at elevated temperatures. For example, σ phase formation is often observed in Cr-containing HEAs. Age hardening of $CuCr_2Fe_2NiMn$ alloy leads to the precipitation of ρ phase which contributes to high-temperature strength of the alloy (Ren et al., 2012). Also, the microstructure present often changes at high temperatures. For example, spherical nanoprecipitates are observed in as-cast $Al_{0.3}CoCrFeNi$ alloy, which change to platelets when the alloy is aged at 700°C and to micro-sized rod-shaped precipitates when aging is done at 900°C (Shun and Du, 2009).

HEAs often have higher resistance to recovery, recrystallization, and grain growth. Tsai et al. (2009b) studied the behavior of deformation and annealing of $Al_{0.5}CoCrCuFeNi$ alloy. They found that significant work hardening occurred during forging even at 900°C, which indicates its low dynamic recovery. In addition, at least 5 h at 900°C were required to soften the sample to the hardness level of the

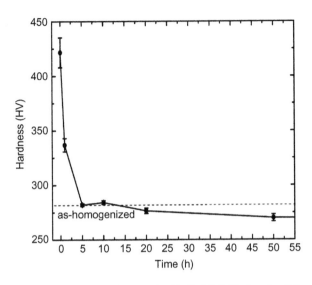

Figure 6.16 Variation of hardness with annealing time for $Al_{0.5}CoCrCuFeNi$ alloy after 50% cold reduction and annealed at 900° C (Tsai et al., 2009b).

as-homogenized state before cold rolling, as shown in Figure 6.16. Since the melting range of the present alloy is from 1279°C to 1362°C, its recrystallization temperature for 1 h annealing is estimated to be 545°C ($\sim 0.5T_m$) according to the empirical rule for traditional alloys. Hence, the real recrystallization temperature of the alloy is much higher than the estimated one by at least 355°C. Bhattacharjee et al. (2014) studied the deformation and annealing behavior of CoCrFeMnNi alloy. They not only found that significantly higher recrystallization temperature than conventional alloy with similar melting point, but also found large resistance to grain growth up to 800°C as shown in Figure 6.4A. They attributed all the above phenomena to solute distorted matrices of HEAs, with high lattice distortion energy, sluggish diffusion effect, and low-stacking fault energy.

CHAPTER 7

Intermetallics, Interstitial Compounds and Metallic Glasses in High-Entropy Alloys

7.1 INTRODUCTION

Research in the field of HEAs for the past one decade has demonstrated that configurational entropy alone is not sufficient to explain the formation of simple random solid solutions in cases where the atomic size differences between atoms are large or when there is a strong attraction between certain elements in the multicomponent alloy. In such cases, various intermetallic phases (or intermediate phases) and under some processing conditions even amorphous phases have been observed. Among the intermetallic phases, the crystal structures most commonly observed are B2, sigma (σ) phase, and Laves phases. Often some of these phases are observed together with random (or disordered) solid solutions and are thought to be due to precipitation from the solid solutions in HEAs during slow cooling or thermal annealing of these alloys. There have been reports on high-entropy nitride and carbide thin films deposited on substrates when the sputtering of HEA targets was subjected to nitrogen and carbon containing atmosphere. They are in fact solid solutions with the structures of interstitial compounds of Hagg phase type. This chapter deals with the various compound structure and amorphous phases formed in HEAs. Nevertheless, it can be said that high-configurational entropy does play a significant role in phase evolution in HEAs, and the number of phases observed in these alloys is significantly less than the maximum number of phases predicted by Gibbs phase rule.

7.2 INTERMETALLIC COMPOUNDS

7.2.1 B2 Phase

The B2 phase is an ordered structure based on BCC (Pearson symbol of cP2), wherein the body-centered position is occupied by one type of atom and the body corners are occupied by atoms of another kind. The most common compounds that have this structure are CsCl, CuZn, and NiAl.

The B2 phase has been observed either as a major or as a minor phase in a large number of HEAs. In some cases, it has been observed to precipitate during heat treatment of these alloys from a BCC phase. In almost all the cases where the B2 phase has been observed, the alloys contained 3d transition elements such as Ti, Cr, Mn, Fe, Co, Ni, and Cu along with Al. Among these 3d transition elements, Al has strong affinity to form B2 phase with Fe, Co, and Ni. A careful observation of the constituents of alloys that showed B2 phase indicates that all have either one of these three elements (Fe, Co, and Ni) along with Al. Not even one alloy has been reported so far in HEA literature that has shown B2 phase without the presence of the above combination of these elements. Thus, the B2 phase that is observed in these alloys can be attributed to the interaction of Al with one of the three elements, while the other elements which also have strong bonding with Al dissolve into the B2 phase mainly due to mixing entropy effect. However, due to the presence of a large number of elements, the order parameter is expected to be low and in a number of cases, the intensity of the superlattice reflections is extremely low, making the researchers to assume the phase to be a BCC solid solution (A2 structure), instead of treating it as an ordered B2 structure.

Alloys containing Cu in addition to B2 phase forming elements have shown Cu-rich FCC phase in their microstructure (Zhang et al., 2010b; Hsu et al., 2007). Studies on $Al_{0.5}CoCrCuFeNiTi_x$ alloy with different x values (Chen et al., 2006b) indicated that at $x = 0.4$, two BCC phases ($\beta1$ and $\beta2$) form, out of which $\beta1$ becomes ordered. Between $x = 0.8$ and 1.2, CoCr-like phase is formed, while Ti_2Ni precipitates when x is in the range of 1.22 in this alloy. The microstructure of the alloy is also affected by the amount of Ti present. Higher Ti content causes Cu to segregate more in ID region because of its more negative enthalpy of mixing with other elements and consequently the area of fraction of ID also increases. The precipitates present in ID region are needle like when Ti content is low while they appear as nanoparticles at higher Ti content (Figure 7.1).

Interestingly, there are differences in phase formation for the same composition in different reports. For example, AlCoCrCuNi prepared by arc melting and casting showed the B2 phase as a major phase along with an FCC phase as a minor phase (Hsu et al., 2007). In contrast, the same alloy prepared by the same technique by Munitz

Figure 7.1 SEM microstructure of as-cast $Al_{0.5}CoCrCuFeNiTi_x$ alloy for (A) x = 0.4 and (B) x = 0.6 (Chen et al., 2006b).

et al. (2013) showed a BCC phase as a major phase and B2 as a minor phase. In AlCoCrFeNi alloy, a single-phase B2 was observed in many cases (arc melting by Wang et al., 2008; Bridgeman solidification by Zhang et al., 2012b; suction casting by Qiao et al., 2011 and Hsu et al., 2013a; and electro-spark deposition by Li et al., 2013c). Thus, the phase formation in HEAs appears to be quite sensitive to the processing conditions. Table A2.1 of Appendix 2 lists the HEAs in which a B2 phase has been observed, either as a single phase or together with some other phase.

7.2.2 L1$_2$ Phase

The L1$_2$ phase is an ordered structure based on FCC (Pearson symbol of cP4), wherein the face-centered positions are occupied by one type of atom and the body corners are occupied by another kind of atom. The most common compounds that have this structure are $AuCu_3$ and Ni_3Al.

A small number of HEAs have shown the presence of L1$_2$ phase (Table A2.2 of Appendix 2). All these alloys containing both Al and Ni possess FCC phase and also show L1$_2$ phase in the matrix (Li et al., 2008b; Hemphill et al., 2012; Manzoni et al., 2013a). As mentioned above, Al content should not be high, otherwise BCC and B2 phases would form. In such HEAs, the L1$_2$ phase is the multicomponent version (including Ni, Co, Fe, Ti, etc.) of γ' Ni_3Al that is usually observed in Ni-base superalloys.

7.2.3 Sigma Phase

The σ phase is usually observed in Cr-containing steels and has a typical composition of equiatomic FeCr with a tetragonal structure. The σ phase has also been observed with equiatomic CoCr or FeMo in binary Co–Cr and Fe–Mo alloys.

A large number of HEAs containing Fe and/or Co together with higher amounts of Cr and/or Mo have shown the formation of σ phase at various stages of their processing. In HEAs, the σ phase is also a multicomponent solid solution. The formation of σ phase is an indication that different types of solid solutions in HEAs could form depending on the interaction and atomic size difference between elements and not just the configurational entropy alone. The σ phase is in fact a topologically close-packed phase in which components with larger atomic size occupy one specific set of lattice sites while smaller atoms occupy another set so as to get a higher number of bonding to lower its overall free energy, although their interactions (or enthalpy of mixing) between components are small. Table A2.3 of Appendix 2 gives the HEAs in which the σ phase has been observed.

Hsu et al. (2010b) studied $AlCo_xCrFeMo_{0.5}Ni$ alloys with different cobalt content. For $x = 0.5$, 1.0, and 1.5, ordered BCC and σ phases are present, whereas a FCC phase appears when $x = 2$ indicating the FCC stabilizing nature of cobalt. The σ phase is believed to be a solution of CoCr, FeCr, FeMo, and NiMo phases. The microstructure of the alloys consists of DR, ID, and eutectic structure at $x = 0.5$; DR and ID structure (without eutectic structure) at $x = 1$ and 1.5; and polycrystalline single phase structure at $x = 2$ (Figure 7.2). In this figure, modulated structures are revealed in (Al,Ni)-rich BCC DR and (Cr,Mo)-rich BCC ID regions at $x = 1$ and 1.5. They form by phase decomposition from BCC phase into σ phase and ordered BCC (B2) phase during the cooling stage. For $x = 2$, σ phase forms along grain boundaries, and σ phase stringers with B2 + FCC mixture between the stringers are observed in grain interiors.

Addition of molybdenum tends to stabilize the formation of a BCC structure and/or appearance of a σ phase. XRD studies of $Co_{1.5}CrFeNi_{1.5}Ti_{0.5}Mo_x$ alloys showed that σ phase begins to form as Mo content reaches $x = 0.5$ (Chou et al., 2010b). Formation of σ phase has also been reported in $CoCrFeNiMo_x$ alloys with increasing Mo content (Shun et al., 2012a). Increasing Mo content has also been

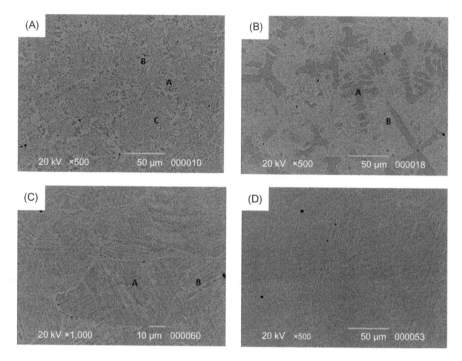

Figure 7.2 SEM backscattered images of as-cast AlCo$_x$CrFeMo$_{0.5}$Ni alloy with (A) x = 0.5, (B) x = 1, (C) x = 1.5, and (D) x = 2. A, DR; B, ID; and C, eutectic.

shown to transform the cast structure to a eutectic structure (Zhu et al., 2010a). When added to AlCrFeNi alloys, Mo dissolves preferentially in the (Fe,Cr)-rich BCC phase while the other BCC phase, based on Al–Ni is lean in Mo (Dong et al., 2013a). Increase in Mo content transforms the eutectic microstructure to hypo/hyper eutectic structure. This preference is probably due to higher driving force for the formation of a stable σ phase of FeCrMo type.

Cr is an important constituent in HEAs, and as already discussed in the context of equiatomic alloys, Cr stabilizes the BCC structure and promotes formation of the σ phase particularly in presence of Fe, Co, and Ni. Hsu et al. (2011) plotted the phase diagram to a first approximation of AlCoCr$_x$FeMo$_{0.5}$Ni alloy (Figure 7.3) for various chromium contents to first approximations based on experimental results of DTA, high-temperature XRD, and microscopic analysis. It indicates that a σ phase is likely to form for all chromium contents along with a B2 phase. At higher temperatures, a FCC phase evolves when the Cr content is less than 20 at.%. It may be noted that the σ phase is also

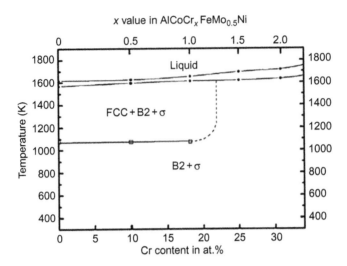

Figure 7.3 Probable phase diagram of AlCoCr$_x$FeMo$_{0.5}$ HEA for different values of Cr (Hsu et al., 2011).

present when $x = 0$; which shows that Cr solely is not responsible for the formation of the σ phase although it certainly accelerates the process. The σ phase can therefore be thought of as a solid solution of various binary σ phases possible like Co–Cr, Fe–Cr, Ni–Mo, Fe–Cr–Mo, Cr–Mo–Ni, and Co–Fe–Mo (ASM Handbook, 1992). Confirmation of such a diagram may prove immensely helpful in establishing the design criteria for various HEAs.

Chen et al. (2006a) discussed the effect of vanadium on FCC Al$_{0.5}$CoCrCuFeNi alloy. The crystal structures of Al$_{0.5}$CoCrCuFeNiV$_x$ alloy system are essentially FCC, BCC, and σ phase, in which the σ phase forms from $x = 0.6$ to 1.0. It is seen that the BCC phase begins to appear at $x = 0.4$ and its amount increases with increase in V content up to $x = 2.0$. The σ phase is a multicomponent solid solution having the same structure as equiatomic NiCoCr phase but this σ phase contains a higher amount of vanadium than other phases in the matrix.

Formation of σ phase is likely to occur when elements like Cr and V are present along with Ni and Co. Chen et al. (2008) reported that σ phase is observed when AlCrCoMoNi layer is produced by TIG cladding. This σ phase has a simple tetragonal structure with lattice parameters of $a = 0.875$ and $c = 0.451$ nm. Chen et al. (2009a) synthesized AlCrCoFeMoNiSi alloy with various Si contents. The Si-free alloy showed the presence of σ phase. With addition of Si, the σ phase disappeared and a new phase FeMoSi evolved which has an hcp structure with lattice parameters $a = 0.765$ and $c = 0.477$ nm.

7.2.4 Laves Phase

Laves phases are intermetallic compounds that have a stoichiometry of AB_2 and are formed when the atomic size ratio is between 1.05 and 1.67. There are three classes of Laves phases, namely, cubic $MgCu_2$ (C15), hexagonal $MgZn_2$ (C14), and hexagonal $MgNi_2$ (C36). In these compounds, the A atoms take up ordered positions as in diamond, hexagonal diamond or related structure while the B atoms take up tetrahedral positions around A atoms. In case the atomic size ratio of A and B atoms is around 1.225, they form topologically tetrahedral close-packed structures with overall packing density of 0.71.

Many HEAs have also shown the formation of Laves phase either as a major phase or a minor phase. Table A2.4 of Appendix 2 lists the HEAs in which Laves phase has been observed. Shun et al. (2012b) have reported the formation of (Ti,Co)-rich Laves phase in $CoCrFeNiTi_{0.5}$ alloy prepared by arc melting. Mishra et al. (2012) have also observed Ti_2Co-type Laves phase in CoCuFeNiTi alloys when Ti/Cu ratio is larger than 9/11. Senkov et al. (2013c) observed Cr-rich Laves phase along with two BCC phases in the arc-melted $CrMo_{0.5}NbTa_{0.5}TiZr$ alloy. In all the above alloys, the formation of Laves phase can be attributed to the presence of Ti along with other transition elements. Dong et al. (2013b) have observed a Laves phase along with FCC and BCC phases when $x = 0.8$ in $Al_xCoCrFeNiTi_{0.5}$. Interestingly, the alloy without Al showed only a FCC phase. Qiu et al. (2014) have observed Laves phase in laser cladded $Al_2CoCrCuFeNiTi_x$ with $x = 0.5$ and 1.5. Ma and Zhang (2012) observed Laves phase in a Nb-containing HEA. They observed (Co,Cr)Nb-type Laves phase along with a BCC phase in $AlCoCrFeNiNb_x$ alloys when $x = 0.25 - 0.75$. At lower Nb content, only a BCC phase has been observed.

In the absence of copper, $Al_{0.5}CoCrFeNiTi_x$ alloys (Zhou et al., 2007b) show that only a single-phase solid solution is present up to $x = 1.0$ and phase separation occurs only at higher Ti contents ($x = 1.5$) with the evolution of a Laves phase of Fe_2Ti type. When Al is not present in the alloy, Ti is shown to precipitate out to form complex phases. $CoCrFeNiTi_x$ alloy which has a single-phase FCC structure when $x = 0$ exhibits presence of R and σ phases when $x = 0.3$ and Laves phase at $x = 0.5$ (Shun et al., 2012b). The R phase is rich in Ni and Ti corresponding to $(Ni,Co,Fe)_2(Ti,Cr)$, the

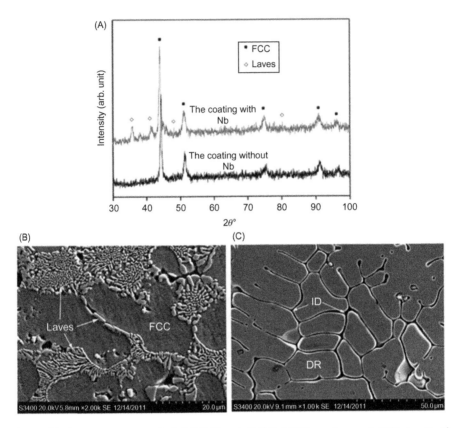

Figure 7.4 (A) XRD patterns of CoCrCuFeNi and CoCrCuFeNiNb coatings and SEM images of (B) CoCrCuFeNiNb and (C) CoCrCuFeNi coatings (Cheng et al., 2014).

σ phase is (Fe,Cr)-rich and of (Cr,Ti)(Fe,Co,Ni)-type, and the Laves phase corresponds to $(Co,Fe,Ni,Cr)_2Ti$ and is rich in Ti and Co.

Kunce et al. (2013) reported the synthesis of CrFeNiTiVZr HEA from elemental powders in an equiatomic ratio using LENS technique. The chemical homogeneity of the alloy was improved through annealing at 1000°C for 24 h. The resulting alloy exhibited a two-phase structure with a dominant C14 Laves phase and a minor α-Ti solid solution. Cheng et al. (2014) developed CoCrCuFeNiNb HEA coating by plasma transferred arc cladding. In the absence of Nb, the alloy showed a single-phase FCC structure in the coating, while the Nb-containing coating showed a (CoCr)Nb-type Laves phase along with a FCC phase (Figure 7.4). The Nb-containing coating showed significant improvement in hardness, Young's modulus, wear resistance, and corrosion resistance in comparison to the coating without Nb.

Table A2.5 of Appendix 2 lists a few HEAs, which show other intermetallic phases including quasicrystalline phase.

7.3 INTERSTITIAL COMPOUNDS (HAGG PHASES)

An interstitial compound is formed when sufficiently small atoms occupy specific interstitial sites in a metal lattice. They are also called Hagg phases. For example, transition metals generally crystallize in either the hexagonal close-packed or face-centered cubic structures providing different sets of tetrahedral and octahedral sites. Their typical stoichiometric compounds are MX, M_2X, MX_2, M_3X and M_6X where M can be Zr, Ti, V, Cr, Fe, etc. and X can be H, B, C, and N.

HEA concept can be extended to create high-entropy ceramics including nitrides, carbides, oxides, and their combinations. Table A2.6 of Appendix 2 lists a few HEAs that have been used to form carbides and nitrides. New interstitial compounds with multicomponent lattice have been synthesized by reactive sputter deposition methods. In 2004, first HE nitride film was deposited from a HEA target of $FeCoNiCrCuAl_{0.5}$ alloy by reactive magnetron sputtering method (Chen et al., 2004). HEA film having a FCC structure and a hardness of 4.4 GPa was obtained without the reaction with nitrogen (i.e., at 0 sccm). As nitrogen flow rate increased, nitrogen content of films increased correspondingly approaching 41 at.% at the flow rate 20 sccm. The structure became a mixture of nanocrystalline HE nitride and an amorphous phase. The hardness was also increased to 10.4 GPa mainly because the number of strong bonding of Al−N, Cr−N, and Fe−N increased.

Based on this, the composition was altered to get higher film hardness by using strong nitride forming elements. Lai et al. (2006b) used the target of equiatomic HEA AlCrTaTiZr to deposit nitride films under different bias voltages in an atmosphere of Ar and N_2. Figure 7.5 shows the XRD patterns of target and films. Although the constituent phases of the target are FCC and a multicomponent phase with the structure of Al_2Zr compound, all the coatings form NaCl-type FCC structure in which N atoms occupy the Cl^- sites and target elements occupy the Na^+ sites. As a result, the HE nitride obtained was a simple solid solution of binary nitrides based on target elements.

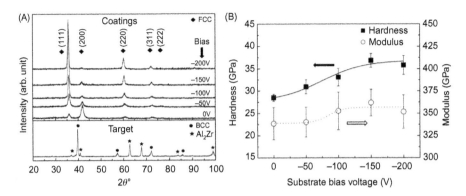

Figure 7.5 (A) XRD patterns of AlCrTaTiZr target and nitride films deposited under different bias voltage and (B) hardness and Young's modulus as functions of bias voltage. Adapted from Lai et al. (2006b).

The composition can be approximated to the stoichiometric formula $(AlCrTaTiCr)_{50}N_{50}$ where the five metal elements are nearly in equiatomic ratio. Hardness and Young's modulus of the coatings increase with increasing bias voltage due to densification, grain refining, and increased residual compressive stress. It is apparent that the hardness level of such HE nitride is much larger than that of HE nitride film deposited from $FeCoNiCrCuAl_{0.5}$. This is mainly because all Al, Cr, Ta, Ti, and Zr can form strong bond with nitrogen.

After this research, more HE nitrides have been developed toward higher hardness, softening resistance, and oxidation resistance in order to prolong the lifetime of cutting tools and dies. Similarly, hard HE carbide films have been researched with reactive sputtering methods by co-sputtering with graphite target or using $Ar + CH_4$ gas flow. Besides the amorphous structure and smaller hardness of $(AlBCrSiTi)N$, other HE nitrides and carbides have NaCl-type FCC structure and much higher hardness and Young's modulus. It has been demonstrated that the FCC structure is thermally stable even after $1100°C$ annealing for 5 h (Tsai et al., 2012; Huang and Yeh, 2010). This suggests that such HE films would have stronger high-entropy effect than the enthalpy effect due to the differences in chemical bonding, atomic size, and crystal structure, and thus avoid phase separation into individual binary or ternary nitrides.

The first high-entropy bulk carbide $(CrNbTiVW)_{50}C_{50}$ with the transition elements in equiatomic ratio was synthesized through MA and solid-state sintering and reported in 2012 (Yeh et al., 2012). The sintered carbide is a thermally stable NaCl-type FCC solid solution of five binary carbides. Nano-powder prepared by ball milling is the key factor to realize the successful pressureless sintering at a low temperature of 1723 K, about 0.56 of the melting point (3108 K) as predicted by the rule of mixture. The average nanohardness value of 32 ± 3 GPa is higher than the rule-of-mixture value of 23 GPa by about 40%, indicating significant solution hardening effect.

7.4 METALLIC GLASSES

As discussed in Chapter 3, the field of metallic glasses is indebted to Pol Duwez and Inoue for their pioneering work in conventional and bulk metallic glasses, respectively. Most of these metallic glasses are based on one principal element (such as Zr-, Fe-, Al-, and Mg-based glasses).

Cantor (2007) has pioneered a new class of glasses by substituting elements of similar nature in equal amounts in ternary glasses. For example, in a Zr−Cu−Al glass forming system, he substituted Ti, Hf and Nb for Zr, Ni and Ag for Cu and could make glassy alloys with about eight elements. The system studied by Cantor's group is (Ti,Zr, Nb,Hf)−(Ni,Cu,Ag)−Al. Though these alloys have as many as eight elements, they can be referred to as pseudo-binary or pseudo-ternary alloys. The configurational entropy of all these alloys is higher than $1.5R$, thus qualifying them to be referred to as HEA glasses. Table A2.7 of Appendix 2 gives a list of multicomponent glassy alloys that have been formed by equiatomic substitution.

In another interesting study, Takeuchi et al. (2011) took the best glass forming composition reported so far ($Pd_{40}Ni_{20}Cu_{20}P_{20}$) and replaced half of Pd with Pt and obtained the first equiatomic metallic glass ($Pd_{20}Pt_{20}Ni_{20}Cu_{20}P_{20}$) by melting the alloy along with a flux in vacuum sealed quartz tube followed by water quenching. However, similar to Cantor's alloys, this alloy can also be referred to as a pseudo-binary alloy as Pt, Ni, and Cu are all similar elements to Pd and substituted for Pd in a $Pd_{80}P_{20}$ alloy. Takeuchi et al. (2013b) have also made CuNiPdTiZr equiatomic glass by melt spinning and in this case it can be considered as equiatomic substitution of Ti for Zr and Ni and Pd for Cu in $Cu_{60}Zr_{40}$ glass which is well known. Similarly, Ding and Yao (2013) took the equiatomic substitution route to form an equiatomic BeCuNiTiZr (or ZrTiCuNiBe) glassy alloy, whose composition can be referred to as $(ZrTi)_{40}(CuNi)_{40}Be_{20}$.

The equiatomic quinary glassy compositions prepared by Gao et al. (2011) by induction melting and subsequently casting into copper mold, namely, CaMgSrYbZn and AlDyErNiTb do not appear to fit into the above equiatomic substitution category and hence can be treated as true equiatomic HEA glasses. However, it is not clear whether the glass formation in these alloys is entropy driven due to the high-configurational entropy of the alloy. In addition to the conventional processing route of melting and casting, several HEA glasses are also obtained either in the thin film form by magnetron sputtering or in the powder form by MA.

Chen et al. (2010b) reported amorphization in BeCoMgTi and BeCoMgTiZn alloys after ball milling for 144 h. This is indicated in Figure 7.6. The amorphization is of Type 2 according to classification of Weeber and Bakker (1988) since no crystalline solid solutions and compounds were formed before full amorphization in BeCoMgTi and

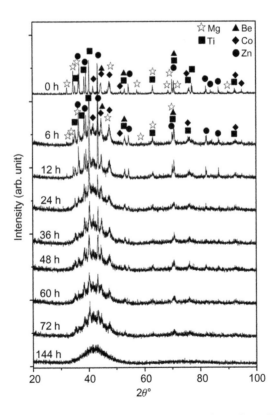

Figure 7.6 Amorphization with prolonged milling in BeCoMgTiZn alloy (Chen et al., 2010b).

BeCoMgTiZn alloys during MA. The inhibition of intermetallic compounds before amorphization is due to chemical compatibility among the constituent elements in conjunction with high-entropy effect and deformation effect which enhance the mutual solubility. Direct formation of the amorphous solid solution phase instead of the crystalline one can be attributed to the large range of atomic size. Table A2.7 of Appendix 2 gives various equiatomic and nonequiatomic HEAs that form glassy structures.

CHAPTER 8

Structural Properties

8.1 INTRODUCTION

HEAs have such promising properties that they are considered as potential candidates for a wide range of applications such as high temperature, electronic, magnetic, anticorrosion, and wear-resistant applications. Many of these properties arise out of their unique structural feature, a multicomponent solid solution. In some cases, HEAs show nanoscale precipitates, which further enhance some of the properties of these alloys. This chapter deals with various structural properties of HEAs including mechanical, wear, electrochemical, and oxidation.

8.2 MECHANICAL PROPERTIES

Mechanical properties cover hardness, elastic modulus, yield strength, ultimate strength, elongation, fatigue, and creep. Structural applications require adequate combinations of these properties. For high-temperature applications, resistance to creep, oxidation and sulfidation (hot corrosion) are taken into account in the material-selection requirements. As a result, studies of mechanical properties of HEAs are important to demonstrate that this new class of alloys have better performance than that of conventional alloys for specific applications. In addition, as HEAs comprise multiprincipal elements and have pronounced core effects as mentioned previously, the mechanisms of structure-property correlations might not be an extension of those based on conventional alloys. Scientific understanding and verification of these correlations becomes an important basic issue.

8.2.1 Room-Temperature Mechanical Properties

Different alloys including refractory HEAs are described in this section. Cast and wrought processes, and powder metallurgy route including mechanical alloying and hot isostatic pressing, are used for synthesis. Hardness, compressive, tensile, fatigue properties, and deformation behavior are discussed.

The first system that has been studied extensively is the $Al_xCoCrCuFeNi$ alloys (Tong et al., 2005a, 2005b; Tung et al., 2007; Tsai et al., 2009b). The hardness of the system increases from 133 HV for $x = 0-0.5$ to 655 HV for $x = 3.0$ (Figure 8.1). This can be attributed to the increase in lattice distortion as Al is the largest atom amongst the constituents of the alloy (Callister, 2003). In addition, Al forms strong bonds with other elements in the alloy as reflected in the enthalpy of mixing (Miedema et al., 1980; Takeuchi and Inoue, 2010). Hence the solid solution strengthening effect increases with higher Al content. It was also observed that as the Al content increases the phase changes from FCC to BCC (including disordered BCC and ordered BCC (B2)). The BCC and B2 phases are stronger than the FCC phase. Apart from this, nanoprecipitates formed due to slow diffusion kinetics also strengthen the material. The SD BCC + B2 modulated structure occurring in the BCC DR regions further strengthens the material (Yeh et al., 2004b; Tong et al., 2005a, 2005b). The cracks due to the indents, whose total length is an indication of toughness, are not present for the compositions where FCC phase is

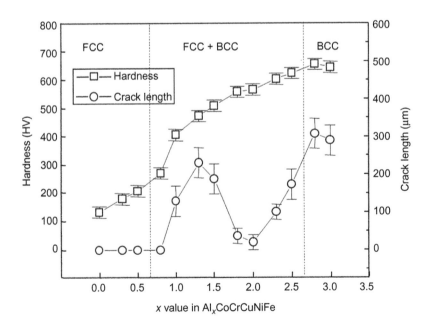

Figure 8.1 Vickers hardness and total crack length around the hardness indent of $Al_xCoCrCuFeNi$ alloy system with different aluminum contents (x values) (Tong et al., 2005a).

dominant (Figure 8.1). But, as the x value increases such that a BCC phase starts to dominate, the cracks start to appear. The length of the crack increases with increased amount of BCC phase, which indicates increased brittleness. Li et al. (2009) reported similar effect of Al on hardness. Addition of V to the $Al_{0.5}CoCrCuFeNi$ alloy to synthesize $Al_{0.5}CoCrCuFeNiV_x$ alloy increases the hardness (Chen et al., 2006a). After $x = 0.4$, the BCC phase becomes the DR phase and increases in volume fraction from thereon. Also a σ phase forms after $x = 0.6$, which further contributes to the hardness of the alloy.

Alloying elements with larger atomic sizes have a tendency to form secondary phases and cause precipitation strengthening. For example, among the 10 systems studied by Li et al. (2009), the highest hardness (566 HV) was achieved with the addition of Zr and Ti, which have larger atomic size and cause precipitation of secondary phases. Ma and Zhang (2012) reported that Nb addition from $x = 0$ to 0.5 increases the hardness of $AlCoCrFeNb_xNi$ alloy from 500 to 750 HV. Combined additions of Ti and Nb increase the hardness to 797 HV, which is higher than that achieved by adding only one of these two elements (Razuan et al., 2013). In another report (Hsu et al., 2010a), $AlCoCrFe_xMo_{0.5}Ni$ alloy's hardness was reported to decrease with an increase in Fe, because the alloy has a soft BCC phase and a harder σ phase in the microstructure. As the amount of Fe increases, the amount of BCC phase increases at the expense of σ phase causing a drop in hardness and wear resistance as

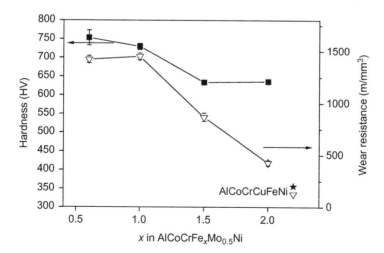

Figure 8.2 Plots of Vickers hardness and wear resistance of $AlCoCrFe_xMo_{0.5}Ni$ alloys as functions of Fe content (Hsu et al., 2010a).

shown in Figure 8.2. Similar to Fe, increase in Co content in $AlCo_xCrFeMo_{0.5}Ni$ decreases the hardness from 788 HV at $x = 0.5$ to 596 HV at $x = 2.0$ (Hsu et al., 2010b).

The $AlCoCrFeNiTi_x$ alloys showed good mechanical properties in compression (Zhou et al., 2007b, 2008a). The composition with $x = 0.5$ had a BCC structure and was found to give optimum properties. The yield strength and fracture strength were found to be 2.26 and 3.14 GPa, respectively, which are much higher than that of BMGs. The alloy also had a 23.3% plastic elongation. The high strength was attributed to the spinodal decomposition in DR region, while there was precipitation of BCC phase particles in the ID region. Thus apart from solid solution strengthening, nanoparticle and precipitation strengthening also play an important role.

The major deformation mechanism in the $Al_{0.5}CoCrCuFeNi$ alloy has been found to be twinning but not slip (Tsai et al., 2009b). This is different from that in conventional FCC alloys in which twinning is not commonly observed. One evidence is that etch pits that were observed in the presence of slip bands were absent in the microstructure and TEM did not reveal dislocation tangles. Figure 8.3 shows the dark-field and bright-field images of the nanotwins formed between the Widmanstätten precipitates and their plane and direction for the 5% cold-rolled alloy (Tsai et al., 2009b). This suggests that nanotwinning in this HEA starts in the initial stage of plastic deformation itself. This is because there are nanoscale precipitates distributed throughout the matrix of DR and slip

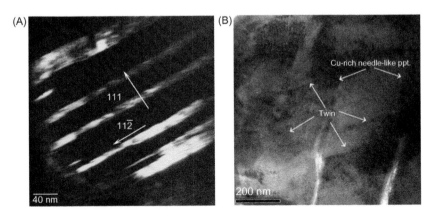

Figure 8.3 (A) Dark-field image and (B) bright-field image of nanotwins between Widmanstätten precipitates in $Al_{0.5}CoCrCuFeNi$ 5% cold-rolled alloy (Tsai et al., 2009b).

movement is easily arrested. One more reason is that presence of large number of elements in HEA alloys causes Suzuki interaction, which significantly reduces the stacking fault energy (SFE), thus making it easier for twinning. Hence, nanotwinning occurs because the twin nucleation is initiated at nanoscale region between the Widmanstätten precipitates. Similar nanotwins have also been observed in CoCrFeNi HEA obtained by mechanical alloying followed by SPS. AlCoCrCuFeNi alloy made by splat quenching and casting was compared by Singh et al. (2011b). The hardness values of splat-quenched (539 HV) and as-cast (534 HV) alloys are almost equal. However, the indentation elastic modulus of the as-cast material (182 GPa) is markedly higher than that of the splat-quenched alloy (105 GPa).

In a single-phase FCC CoCrFeMnNi alloy (Otto et al., 2013a), it was found that the deformation is governed by planar slip of $1/2\langle 110\rangle$ type dislocations on $\langle 111\rangle$ planes at strains $<2.4\%$, which also indicates low SFE. At a strain above 20%, the prevalence of deformation twinning was observed at 77 K, but the formation of dislocation cell structures was observed at 293 K. The evolution of microstructure and texture after heavy cold rolling and annealing in the same CoCrFeMnNi HEA was also reported (Bhattacharjee et al., 2014). A predominantly brass-type texture after 90% cold rolling was observed indicating the low SFE of this alloy. They proposed that the combined effect of high energy level of the distorted matrix and the strain energy relief of stacking fault by *in situ* atom position adjustment results in very low SFE as compared with conventional alloys featured by the small-distorted matrix with a host element.

Using refractory elements (melting points higher than 1650°C) as constituents, the American Air Force Research Laboratory developed refractory HEAs in order to overcome the temperature limitation associated with conventional superalloys. The room-temperature compressive yield stress of the first reported alloys MoNbTaW and MoNbTaVW was found to be 1058 and 1246 MPa, respectively. The quaternary MoNbTaW alloy showed a maximum strength of 1211 MPa and failed after 2.1% strain by a splitting mechanism (Senkov et al., 2010, 2011b). The Young's modulus of this alloy was 220 ± 20 GPa. On the other hand, the quinary alloy MoNbTaVW showed a maximum strength of 1270 MPa and failed after 1.7% strain by the same mechanism. The Young's modulus of this alloy was

180 ± 15 GPa. The failure occurred in a quasi-cleavage fashion parallel to the stress direction and hence in tensile mode.

Another refractory HEA HfNbTaTiZr, was also synthesized by arc melting (Senkov et al., 2011a). This alloy has a slightly lower yield stress of 929 MPa than the two refractory HEAs discussed above but deforms in compression up to 50% strain without any sign of fracture. This alloy showed good work hardening at a steady rate. Figure 8.4 shows uniform deformation behavior and evidences of twins and flow lines. Thus, the alloy can be strained to required strength and still has good ductility. Because most BCC alloys and ordered BCC (B2-type) intermetallic phases are brittle and tend to fracture by cleavage, this finding provides a good basis for developing ductile BCC alloys by modifying the composition.

Most of the studies measure only compressive properties. Yet, there are a few studies done exclusively to find out the tensile behavior of HEAs. Tensile behavior of the alloy $Al_{0.5}CoCrCuFeNi$ was studied on the cold-rolled sample with 80% reduction (Tsai et al., 2010a). At room temperature, the alloy showed yield strength of 1292 MPa, ultimate tensile strength of 1406 MPa, and elongation of 6%. Thus, this alloy has a combination of strength and ductility better than cold-rolled 304 stainless steels. These cold-rolled samples were further annealed at 900°C and their tensile properties were studied. The combination of yield strength (656 MPa), ultimate tensile strength (796 MPa), and elongation (29%) of this alloy is better than that of 304 stainless steel (310 and 620 MPa, and 30%, respectively). The alloy

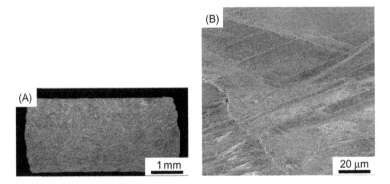

Figure 8.4 (A) HfNbTaTiZr showing uniform deformation following compression and (B) corresponding BSE (back-scattered electron) micrograph following compressive deformation (Senkov et al., 2011a).

in the as-annealed state thus has remarkable mechanical properties for engineering applications.

Hemphill et al. (2012) were the first to study the fatigue behavior of an $Al_{0.5}CoCrCuFeNi$ HEA. The arc-melted samples were annealed at 1000°C for 6 h, water quenched, and cold rolled. The rolled sheets were subsequently machined to fatigue samples for four-point bending fatigue test. The fatigue investigation indicates that the fatigue behavior of HEAs compares favorably with that of many conventional alloys, such as steels, titanium alloys, and advanced BMGs with a fatigue endurance limit of between 540 and 945 MPa and a fatigue endurance limit to ultimate tensile strength ratio of between 0.402 and 0.703. The encouraging fatigue resistance demonstrates that $Al_{0.5}CoCrCuFeNi$ HEAs without surface defects may be useful in future applications where fatigue is a factor.

All the above studies on room-temperature mechanical properties are carried out on cast or wrought alloys. As discussed previously, when the alloys are manufactured by MA and sintering, usually better strength and hardness are achieved. The nanocrystalline nature of the structure causes the increase in hardness along with solid solution strengthening. Hardness of AlCrCuFeTiZn solid solution was 2 GPa in the MA-sintered condition. The same alloy when processed using MA-vacuum hot pressing (VHP) had a hardness of 9.50 GPa and compressive strength of 2.19 GPa and those after MA-hot isostatic pressing (HIP) were 10.04 and 2.83 GPa, respectively. Wear resistance of the alloy was found to be higher than that of the commercially used materials such as Ni-hard-faced alloy. Similarly, hardness and compressive strength of the AlCoCuNiTiZn HEA processed by MA-VHP were found to be 7.55 and 2.36 GPa, respectively. When HIP is used for the same alloys, hardness and compressive strength were 8.79 and 2.76 GPa, respectively (Varalakshmi et al., 2008, 2010a–c).

$Al_xCoCrCuFeNi$ ($x = 0.45$, 1, 2.5, and 5) alloys synthesized by MA and SPS were studied by Sriharitha et al. (2014). Highest specific hardness of 960 HV was achieved in the sintered $Al_5CoCrCuFeNi$ alloy. Hall–Petch analysis based on hardness measurements carried out on sintered samples reveals that the contribution of solid solution strengthening and order strengthening, in comparison to grain size strengthening, increases with the increase in Al content (Figure 8.5).

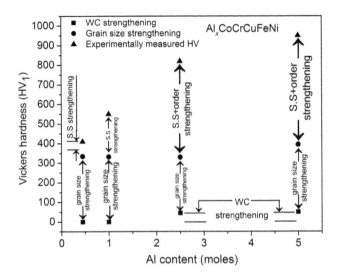

Figure 8.5 Vickers hardness and Hall—Petch analysis of sintered Al$_x$CoCrCuFeNi (x = 0.45, 1, 2.5, and 5) alloys. S.S in the figure indicates solid solution (Sriharitha et al., 2014).

Oxide-dispersed AlCoCrFe HEA synthesized by MA and SPS was studied by Praveen et al. (2013a). High hardness values of 1050 and 1070 HV were observed without and with oxide dispersion, respectively. It appears that solid solution strengthening effect in HEAs has superseded the effect of oxide dispersion. A hardness of 570 HV was achieved in CoCrFeNi alloy by the same synthesis technique (Praveen et al., 2013b).

8.2.2 High-Temperature Mechanical Properties

Due to sluggish diffusion effect and second-phase strengthening, HEAs might exhibit high strength at elevated temperatures. For example, AlCo$_x$CrFeMo$_{0.5}$Ni alloy (x = 1) shows a hardness of 347 HV at 1273 K, which is higher than that of Ni-based superalloys IN 718/IN 718H by 220 HV (Hsu et al., 2010b). AlCoCrFeMo$_{0.5}$Ni$_x$ (x = 0—1.5) also shows that hot hardness remains higher than that of superalloys IN 718 and IN 718H up to 1273 K (Figure 8.6). These alloys also possess a less negative value of softening coefficient (B$_{II}$) at high-temperature regime compared with the same superalloys (Juan et al., 2013). Besides, AlCoCuNiTiZn alloy produced by MA-VHP was found to be stable at elevated temperature of about 800°C as it retained its nanostructure (Varalakshmi et al., 2010c).

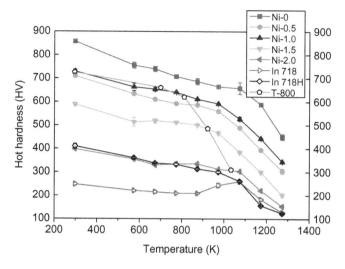

Figure 8.6 Hot hardness versus temperature plots for AlCoCrFeMo₀.₅Niₓ alloys with varying Ni content (Juan et al., 2013).

MoNbTaW and MoNbTaVW HEAs exhibit high yield strength of 405 and 477 MPa, respectively, at 1600°C which is higher than the melting point of superalloys (Senkov et al., 2010, 2011a). Senkov et al. (2013a, 2013b) also developed notable low-density refractory HEAs including NbTiVZr, NbTiV₂Zr, CrNbTiZr, and CrNbTiVZr having densities, 6.52, 6.34, 6.67, and 6.57 g/cc, respectively. Figure 8.7 compares the compression curves of these alloys from 298 to 1273 K, some of which reached 50% strain without fracturing. It is interesting to note that although the room-temperature ductility of Cr-containing HEAs is somewhat low, their high-temperature ductility is remarkably high. In addition, the specific strength of CrNbTiVZr alloys is far superior than that of the other three HEA alloys and compared to In718 and Haynes 230 superalloys. Therefore, CrNbTiVZr alloy shows the most attractive properties, such as considerably improved elevated-temperature strength, reduced density, and much higher melting point. Based on this, better combination of strength and ductility with low density could be expected if compositions and microstructure are further adjusted. As a result, Senkov et al. (2013a, 2013b) recommended a reasonable approach for the development of HEAs with both solid solution and ordered phases as candidates for the next generation of high-temperature structural materials.

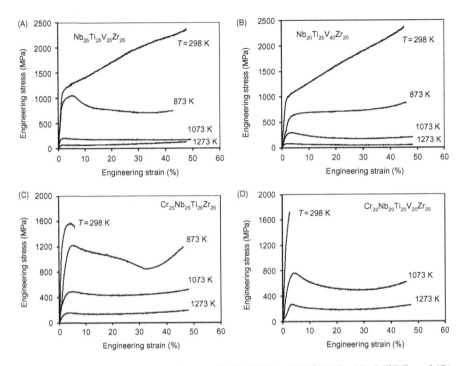

Figure 8.7 Engineering stress versus strain curves of (A) NbTiVZr, (B) NbTiV₂Zr, (C) CrNbTiZr, and (D) CrNbTiVZr alloys at T = 298, 873, 1073, and 1273 K (Senkov et al., 2013a).

Superplastic property was also studied for AlCoCrCuFeNi HEA (Kuznetsov et al., 2013). A fine equiaxed duplex structure with an average grain size of ~1.5 µm was formed after hot multidirectional forging. The forged alloy exhibited superplastic behavior in the temperature range of 800−1000°C at a strain rate of 10^{-3}/s. Elongation to failure approached 604% at 800°C, decreased to 405% at 900°C, and increased again to 860% at 1000°C. Under the strain rate from 10^{-4} to 10^{-2}/s at 1000°C, the superplastic elongation was in the range of 753−864%. An increase in the strain rate to 10^{-1}/s still gave an elongation of 442%. This study demonstrates that HEAs also have the potential in obtaining high-strain rate superplasticity at strain rates higher than 10^{-2}/s.

8.3 WEAR PROPERTIES

Wear resistance and tribology behavior are important for the applications with moving counterparts with/without incorporated abrasive particles, which might be at high temperatures, under impact loading, and/or in

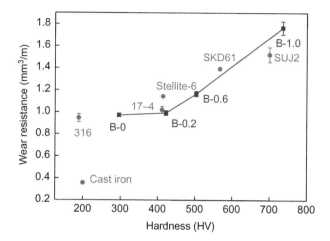

Figure 8.8 Wear resistance as a function of hardness for Al$_{0.5}$CoCrCuFeNiB$_x$ (x = 0, 0.2, and 0.6) HEAs and typical conventional alloys (Hsu et al., 2004).

corrosive environment. In general, hard, tough, lubricating, thermally stable, and chemically non-reactive materials are ideal for such applications. To develop such materials for use in severe operating conditions is really a challenge. Conventional alloys still encounter many bottlenecks in this regard.

Wear properties have been studied from the early stage of developing HEAs, although data are still limited. Hsu et al. (2004) studied abrasion wear resistance of Al$_{0.5}$B$_x$CoCrCuFeNi ($x = 0$, 0.2, 0.6, and 1) HEAs and found that the volume fraction of (Fe,Cr)-rich boride increases with B content. Figure 8.8 compares their wear resistance with that of typical wear-resistant alloys. The alloy with $x = 1$ has better wear resistance than SUJ2 bearing steels. V and Ti additions to the same alloy were also reported (Chen et al., 2006a, 2006b). The wear resistance was rapidly improved with increasing the Ti content from 0.6 to 1.0 and reached a maximum at $x = 1.0$ followed by gradual decrease at higher Ti contents. The optimum is due to the formation of CoCr-like phase between $x = 0.8$ and 1.0. The wear resistance increases by around 20% as the content of V increases from $x = 0.6$ to 1.2 and levels off beyond $x = 1.2$. This improvement is mainly due to the formation of a very hard σ phase.

Tong et al. (2005a) reported the abrasion wear resistance of Al$_x$CoCrCuFeNi ($x = 0.5$, 1.0, 2.0). They found that Al$_{0.5}$ alloy having a FCC structure and a hardness of 223 HV displays very high wear resistance close to that of SKD-61 with a hardness of 567 HV. $Al_{1.0}$ alloy and $Al_{2.0}$ alloy do not have better wear resistance even though their hardness values are around 410 and 570 HV, respectively. The excellent wear resistance of $Al_{0.5}$ alloy is consistent with its high work hardening and ductility, which causes surface to be hardened by deformation during abrasion. Tsai et al. (2009b) had verified by TEM analysis that the large work hardening and ductility of $Al_{0.5}$ alloy are related to the formation of nanotwins due to its low SFE. The other two BCC phase-containing alloys with higher Al contents show wear resistance no better than $Al_{0.5}$ alloy because their BCC phase has a hardness of only around 600 HV and is brittle by nature.

Chuang et al. (2011) reported excellent adhesion wear resistance of $Al_{0.2}Co_{1.5}CrFeNi_{1.5}Ti$ HEA, which has a hardness of 717 HV and a resistance 3.6 times that of SUJ2 (AISI 52100) with similar hardness. In addition, the HEA also displays a wear resistance twice that of high-speed tool steel SKH51 (AISI M2) with a hardness of 870 HV. They demonstrated that this outstanding performance is due to its superior oxidation resistance and remarkable hot hardness over that of comparable steels because the contact temperature at the pin-disk interface can be as high as 800°C.

Plasma nitriding at 525°C for 45 h under a gas mixture of 25% $N_2 + 75\%H_2$ has been applied to improve abrasive wear resistance of HEAs (Tang et al., 2009, 2010). For example, $Al_{0.3}CrFe_{1.5}MnNi_{0.5}$ alloys after different processing routes can be well nitrided, with a thickness of around 80 μm, to attain a peak hardness level around 1300 HV near the surface. The main nitride phases in the surface layer are CrN, AlN, and $(Mn,Fe)_4N$. Nitrided $Al_{0.3}CrFe_{1.5}MnNi_{0.5}$ alloys have much better wear resistance than unnitrided ones by 49 to 80 times and also better than that of conventional nitrided alloys, nitriding steel SACM-645 (AISI 7140), 316 stainless steel, and hot-mold steel SKD-61 (AISI H13) by 22 to 55 times. The superiority is because the nitrided HEA alloys possess a much thicker highly hardened layer than the conventional alloys.

8.4 ELECTROCHEMICAL PROPERTIES

In principle, the overall composition and microstructure of an alloy affect its corrosion resistance in different corrosive environments.

Some HEAs have demonstrated excellent performance in both H_2SO_4 and NaCl solutions. Similar to conventional alloys, it is interesting to note that Cr, Ni, Co, Ti in HEAs enhance corrosion resistance in acid solutions, Mo tends to inhibit pitting corrosion, whereas Al and Mn display a negative effect. However, detailed investigation of the mechanisms is still required.

Figure 8.9 plots the potentiodynamic polarization curves of $AlCoCrCu_{0.5}FeNiSi$ alloy and 304 stainless steel in 0.1 M NaCl solution (Chen et al., 2005d, 2005b, 2005c, 2006c). In this solution, the HEA displays passivation behavior. Corrosion potential is higher than that of 304 steel, and corrosion current density is smaller than that of 304 stainless steel. Similar superiority is observed in 1 M NaCl solution. On the other hand, Table 8.1 gives the average corrosion rates (in mpy) obtained from polarization curves and immersion tests of as-cast $CoCrCu_xFeNi$ alloy ($x = 0$, 0.5, 1) in 3.5% NaCl solution (Hsu et al., 2005). These data demonstrate that Cu addition is detrimental to pitting resistance. However, CoCrFeNi is still better than 304L stainless steel in pitting resistance.

Potentiodynamic polarization studies of as-cast $Al_{0.5}CoCrCuFeNi$ alloy and 304 stainless steel in 1N H_2SO_4 solution (Lee et al., 2007)

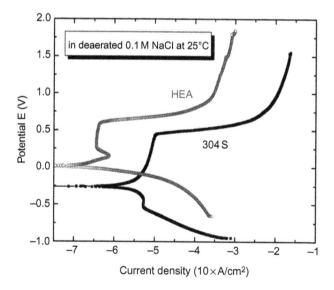

Figure 8.9 Potentiodynamic polarization curves of AlCoCrCu₀.₅FeNiSi alloy and 304 stainless steel in 0.1 M NaCl solution (Chen et al., 2005b).

Table 8.1 Average Corrosion Rate of CoCrCu$_x$FeNi HEAs Obtained from Immersion Test and Polarization Curve in 3.5% NaCl Solution (Hsu et al., 2005)

	CoCrFeNi	CoCrCu$_{0.5}$FeNi	CoCrCuFeNi
Immersion test	7.62×10^{-4}	8.89×10^{-3}	1.14×10^{-2}
Polarization test	3.31×10^{-4}	7.46×10^{-3}	1.37×10^{-2}

indicate that corrosion potential of HEA is higher than that of 304 stainless steel and corrosion current density is smaller than 304 stainless steel, indicating general corrosion is better than that of 304 stainless steel. However, it is clear that HEA has poorer passivation behavior. This can be attributed to the formation of Cu-rich ID phase which provides local galvanic cell to enhance pitting. Similar studies have been carried out on as-cast Al$_x$CrFe$_{1.5}$MnNi$_{0.5}$ alloys in 0.5M H$_2$SO$_4$ solution (Lee et al., 2008b). They all exhibit good passivation regions. However, Al-free alloy has better corrosion resistance than Al-containing alloys because it has smaller passive current density and corrosion current density.

The electrochemical characteristics of Co$_{1.5}$CrFeMo$_x$Ni$_{1.5}$Ti$_{0.5}$ ($x = 0$, 0.1, 0.5, 0.8) alloys in solutions of 0.5M H$_2$SO$_4$ and 1 M NaCl have been studied (Chou et al., 2010a, 2010b, 2011). The potentiodynamic polarization curves of the HEAs in acidic solution exhibit active−passive corrosion behavior, yielding an extensive passive region. This indicates that Mo-free alloy is more resistant to general corrosion than that of the Mo-containing alloys in acidic environments. On the other hand, the cyclic polarization curves demonstrate that the Mo-free alloy is susceptible to pitting corrosion in 1 M NaCl solution, but Mo-containing alloys are not and also their passive films can be repaired by themselves.

8.5 OXIDATION BEHAVIOR

In conventional alloys, oxidation resistance could be largely improved by adding suitable amounts of Al, Cr, and Si, because these elements could form dense and stable oxide layer on the surface at high temperatures (Sims and Hagel, 1972). By the same principle, many HEAs containing such elements generally display improved oxidation resistance. Thus, AlCoCrFeNi and AlCoCrFeMoNi HEAs often show good oxidation resistance up to 1100°C. AlCrFeMnNi alloy often has less oxidation resistance when Mn content is higher.

The oxidation resistance of refractory HEAs is quite poor because refractory elements such as Ti, Zr, and Hf have a strong affinity to oxygen but their oxides are not adherent by nature. In addition, oxide of V has a low melting point, and those of Mo and W have low boiling points. Senkov et al. (2012b) investigated the isothermal oxidation behavior of a refractory HEA $CrMo_{0.5}NbTa_{0.5}TiZr$ during heating at 1273 K for 100 h in flowing air. Continuous weight gain occurred during oxidation, and the time dependence of the weight gain per unit surface area can be described by a parabolic dependence with the time exponent $n = 0.6$. The alloy has a better combination of mechanical properties and oxidation resistance than commercial Nb alloys and earlier reported developmental Nb−Si−Al−Ti and Nb−Si−Mo alloys. Liu et al. (2014) studied four types of new refractory HEAs $Al_{0.5}CrMoNbTi$ (H−Ti), $Al_{0.5}CrMoNbV$ (H−V), $Al_{0.5}CrMoNbTiV$ (H−TiV), and $Al_{0.5}CrMoNbSi_{0.3}TiV$ (H−TiVSi$_{0.3}$). As expected, these refractory HEAs mainly consist of a simple BCC solid solution due to the high mixing entropy effect. But, the oxidation kinetics of all the refractory HEAs at 1300°C follows a linear behavior although there are some differences in oxidation rate among them. The oxidation resistance of the HEAs is significantly improved with Ti and Si addition, but reduced with V addition. Thus, improvement of oxidation resistance of refractory HEAs is still a very important issue for high temperature applications as compared to superalloys. In another study, Zhu et al. (2014) prepared fully dense Ti(C,N) cermets with AlCoCrFeNi HEA as a binder. They have demonstrated that the HEA binder gives much better oxidation resistance than the conventional Ni/Co binder (Figure 8.10).

Two alloy powders of $AlCoCrFeMo_{0.5}NiSiTi$ and $AlCrFeMo_{0.5}NiSiTi$ have been plasma sprayed on alumina substrate as coating layer around 160 μm in thickness (Huang et al., 2004). Their oxidation resistance was measured by weight gain after exposure to oxygen at high temperatures. From the atmospheric oxidation experiment, both coating layers possessed good oxidation resistance at 1000°C, and even up to 1100°C since their weight gains in the formation of oxides approached a constant level after

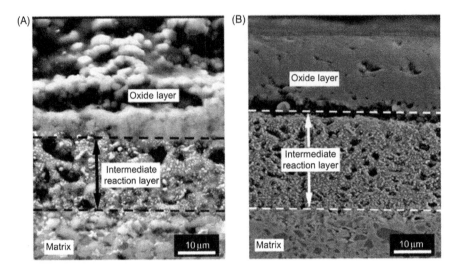

Figure 8.10 Cross-sectional SEM images of Ti(C,N) cermet after isothermal oxidation at 1100°C for 4 h in static air with (A) Ni/Co and (B) AlCoCrFeNi HEA as binders (Zhu et al., 2014).

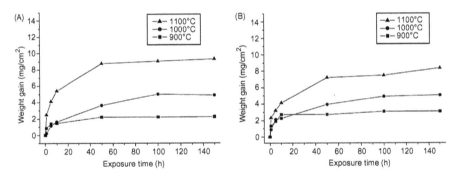

Figure 8.11 Oxidation kinetics of (A) AlSiTiCrFeNiCoMo$_{0.5}$ and (B) AlSiTiCrFeNiMo$_{0.5}$ plasma-sprayed coatings. Adapted from Huang et al. (2004).

about 50 h as shown in Figure 8.11. Passive oxide layers were formed and protected the underlying materials as evident from EDS mapping. The top protective layer was of Ti oxide and next layer was of Cr oxide. Since Ti oxide could not provide a good protection, it is believed that the protection to oxidation is mainly attributable to Cr.

CHAPTER 9

Functional Properties

9.1 INTRODUCTION

Many studies on HEAs are focused on the relationships between microstructure and mechanical properties. A few investigations are reported on the functional properties of these alloys. These include behaviour of HEAs as diffusion barrier, electrical, thermal, magnetic, hydrogen storage, irradiation resistance, and catalytic materials. These properties are also affected by their unique structural feature, that is, a multicomponent solid solution. The properties reported are encouraging and promising for different functional applications. Attempts are being made to take advantage of the encouraging functional properties of these HEAs in applications such as biomedical, antibacteria, electromagnetic interference (EMI) shielding, antifingerprint, antisticky, and hydrophilic and hydrophobic properties.

9.2 DIFFUSION BARRIER PROPERTIES

One of the major challenges in the miniaturization of modern microelectronic devices is to develop future high-performance diffusion barrier materials (International Technology Roadmap for Semiconductors, 2009). When Cu is incorporated into Si, adverse effects including the formation of deep trap levels that cause serious device degradation and failure occur. The metallic barriers between Cu and Si are expected to fail around 550–650°C whereas ceramics fail around 700–800°C. In this aspect, HEA alloy barriers and HEA nitride barriers have shown improved temperature capability (Tsai et al., 2008b,c, 2011; Chang et al., 2009).

Metal barriers fail at lower temperatures due to two main reasons: (1) enhanced diffusion of Cu through barrier grain boundaries; or (2) reaction between barrier metal and Si (Tsai et al., 2011). The first problem can be alleviated by using amorphous ceramics, TaSiN is a classic example. However, in the case of metallic barriers, this strategy is not effective because many amorphous metal barriers crystallize below 550–650°C. Thus, the low structural stability of metals renders them ineffective in preventing the diffusion of Cu. Even those with extremely high melting points, such as W (3407°C), Ta (2996°C), and Mo (2617°C) start to react with Si at around 550−650°C. The second problem of metallic barriers, on the other hand, is due to their low chemical stability and reaction with Si. Thus, a better metal barrier has to maintain its amorphous structure and its inertness to Cu and Si up to temperatures higher than 650°C.

Based on this consideration, Tsai et al. (2008b) designed two suitable HEA alloy compositions for diffusion barrier. The first one is an octonary AlMoNbSiTaTiVZr HEA barrier of 100-nm-thick incorporated into the Cu/HEA/Si sandwich structure. It can prevent interdiffusion of Cu and Si at 700°C up to 30 min. Through further composition design, a 20-nm-thick NbSiTaTiZr diffusion barrier capable of preventing the interdiffusion and reaction between Cu and Si at 800°C up to 30 min was developed as shown in Figure 9.1 (Tsai et al., 2011). These results are comparable to many ceramic barriers. The superior performance of NbSiTaTiZr is owing to the higher structural stability of its amorphous structure and higher chemical stability against reaction with Si substrate as compared to conventional metal barriers. The structural stability comes from the low driving force (0.76 kJ/mol) for crystallization of amorphous phase, the large atomic size difference enhancing the stability of amorphous structure, and the slow diffusion kinetics due to four refractory elements and higher

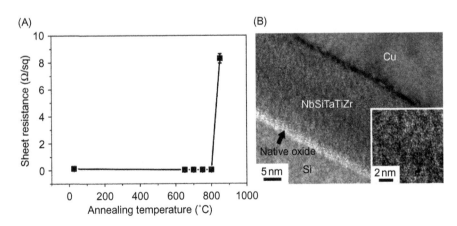

Figure 9.1 (A) Sheet resistance as a function of annealing temperature for the Cu/NbSiTaTiZr/Si test structure and (B) TEM image of the Cu/NbSiTaTiZr/Si test structure after 800°C annealing. Inset shows the high resolution image of the NbSiTaTiZr barrier (Tsai et al., 2011).

atomic packing density in the amorphous structure. The chemical stability comes from its lower free energy level. It was calculated that when unalloyed NbSiTaTiZr reacts with Si to form silicides, the formation enthalpy is 120.5 kJ/mol. However, the driving force for the silicide formation in NbSiTaTiZr is only 76.7 kJ/mol. Thus, the driving force for silicide formation from the alloy is 36% lower than that from pure elements (Tsai et al., 2011).

Tsai et al. (2008a) developed 70-nm-thick nitride diffusion barrier of $(AlMoNbSiTaTiVZr)_{50}N_{50}$ which has an amorphous structure that successfully prevents the reaction between copper and silicon when annealing at 850°C for 30 min. Its amorphous structure remained even after this annealing treatment. The excellent thermal and structural stability for preventing the silicide formation was attributed to the large lattice distortion and limited diffusion. Chang et al. (2009) developed an ultrathin quinary nitride film (AlCrTaTiZr)N of only 10 nm thick as a diffusion barrier layer for Cu interconnects. The (AlCrTaTiZr)N barrier is a nanocomposite constructed of nanocrystallites embedded in an amorphous matrix. At an extremely high temperature of 900°C, the Si/(AlCrTaTiZr) N/Cu film stack remained thermally stable. Neither interdiffusion between Si and Cu through the (AlCrTaTiZr)N layer nor the formation of any silicides was observed. The nanocomposite structure and severe lattice distortions were also discussed as the dominant factors for the superior resistance to diffusion in the (AlCrTaTiZr)N film. In summary, the films of HEAs and their nitrides have shown great promise in this area.

9.3 ELECTRICAL PROPERTIES

Chou et al. (2009) investigated the electrical resistivity of HEAs. The experimental alloys were a series of $Al_xCoCrFeNi$ alloys varied in Al content ($0 \leq x \leq 2$). The effect of Al was investigated because it has a larger atomic size and is a strong BCC former. The arc-melted samples received homogenization treatment at 1100°C for 24 h before investigation.

Figure 9.2 shows the electrical resistivity as a function of temperature and Al content (Chou et al., 2009). The electrical resistivity of each alloy is approximately a linear function of temperature with a

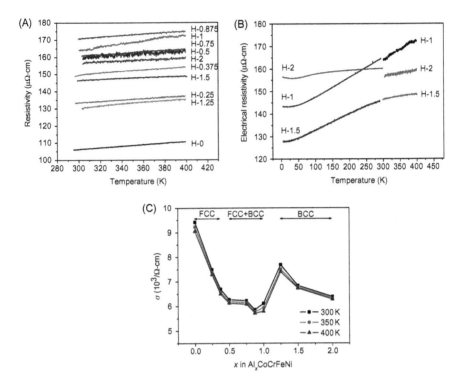

Figure 9.2 Diagrams for (A) electrical resistivity of all alloys in the temperature range of 300–400 K, (B) electrical resistivity of the alloys with x = 1, 1.5, and 2 in the range of 4–400 K. The curves in 4–298 K and those in 298–400 K were measured with two apparatus, and (C) electrical conductivity versus x in Al$_x$CoCrFeNi alloys at 300, 350, and 400 K (Chou et al., 2009).

small positive slope in the range of 298−400 K. At temperatures near absolute zero, the resistivity levels are generally high as compared to that (1−100 μΩ-cm) of conventional alloys. The high-level of electrical resistivity can be attributed to the severe lattice distortion and thus intense electron scattering in these alloys. Furthermore, the thermal vibration effect causing electron scattering becomes relatively smaller as compared with severe-lattice-distortion effect. This makes the alloy to have a low temperature coefficient of resistivity (TCR). This low TCR tendency might be further used as a base to develop new HEAs with even lower TCR. Potential applications in resistors are expected. The variation in electrical conductivity of Al$_x$CoCrFeNi alloy is in three stages and can be related to the constituent phases present in each stage: single-FCC in $0 \leq x \leq 0.375$, single-BCC in $1.25 \leq x \leq 2$, and duplex FCC−BCC in $0.5 \leq x \leq 1$. The electrical conductivity decreases with increasing x in single-phase regions and has a lower level in the duplex phase region. This could be explained based on the fact that larger amount of Al causes more severe lattice distortion and electron scattering, and also the interface between FCC and BCC phase increases electron scattering. Moreover, BCC phase has a higher conductivity than FCC phase.

9.4 THERMAL PROPERTIES

Figure 9.3 shows the thermal conductivity, κ, of HEAs as a function of temperature and Al content (Chou et al., 2009). It is seen that the thermal conductivity increases with temperature, contrary to that of pure metals. For pure metals, the collision frequency of electrons with phonons of the lattice increases with temperature and results in a shorter mean free path and thus a decreased κ. In the present alloys, lattice distortion effect that makes a lower temperature sensitivity of phonon concentration as well as thermal expansion effect that causes larger lattice spacings increase the mean free path of electrons and thus κ with temperature. Figure 9.3B shows κ as a function of x. It can be seen that κ decreases with x in each single-phase region but BCC phase has a higher κ level than FCC phase. The level of κ of alloys with $x \leq 1$ is only around $10-12$ W/mK, which is very low compared with conventional alloys. This is very encouraging for those applications requiring low thermal conductivity. In addition, κ in the FCC + BCC duplex region is at the lowest level for each temperature. This trend is similar to that for electrical conductivity in Figure 9.2C.

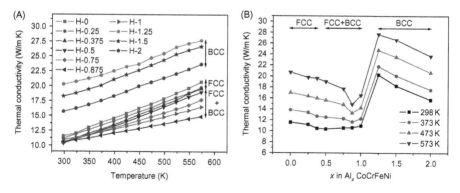

Figure 9.3 (A) Curves of thermal conductivity as a function of temperature and (B) as a function of x *for Al$_x$CoCrFeNi alloys (Chou et al., 2009).*

The explanation for this similarity could be based on the fact that κ has two contributions: $\kappa = \kappa_e + \kappa_{ph}$ where κ_e is from electron conduction and κ_{ph} from phonon conduction. This is because κ_e is proportional to electrical conductivity and thus more sensitive to lattice distortion than phonon conductivity. On the other hand, one more reason for higher κ level of BCC structure is explained on the basis of phonon velocity. That is, higher Young's modulus, E, of the BCC structure gives a higher phonon velocity, $v_{ph} = (E/\rho)^{1/2}$, and thus a greater phonon contribution to κ (Chou et al., 2009).

9.5 MAGNETIC PROPERTIES

Fe, Ni, and Co are the main elements that contribute to ferromagnetism in an alloy. Suitable selection and amount of these elements and addition of other elements to HEAs could produce ferromagnetism and other magnetic properties.

Ma and Zhang (2012) reported that $AlCoCrFeNb_xNi$ ($x = 0$, 0.1, 0.25, 0.5, and 0.75) alloys show ferromagnetic properties. With the increase in Nb content, saturation magnetizations and residual magnetizations decreased. The residual magnetization (M_r) reaches a maximum for $AlCoCrFeNb_{0.1}Ni$ alloy, which is 6.106 emu/g. AlCoCrCuFeNi alloy is reported to possess high saturation magnetization and undergoes a ferromagnetic transition in both as-cast and annealed (2 h at 1000°C) conditions. The values of saturation magnetization (M_s) were observed to be 38.178 and 16.082 emu/g, respectively (Zhang et al., 2010b). Lucas et al. (2011) investigated the magnetic behavior of stoichiometric FeNi, CoCrFeNi, CoCrFeNiPd, $CoCrNiFePd_2$, $Al_2CoCrFeNi$, and AlCoCrCuFeNi alloys. They found that these HEAs based on CoCrFeNi have low saturation magnetization and Curie temperature as compared with Fe and FeNi, and become poor candidates for soft magnetic applications at high temperatures. They attributed this phenomenon to the alloying with antiferromagnetic Cr as reflected by the fact that CoCrFeNi was paramagnetic at room temperature. However, alloying with Pd could increase the magnetic moment and Curie temperature in the FCC phase. They pointed out that the control of the Curie temperature with Pd additions may make these alloys useful for magnetic refrigeration applications near room temperature.

Lucas et al. (2013) also studied the effect of Cr concentration, cold rolling, and subsequent heat treatments on the magnetic properties of $CoCr_xFeNi$ alloys with $x = 0.5-1.15$. CoCrFeNi HEA has a ferromagnetic Curie temperature of 130 K. When Cr content is reduced, an increase in Curie temperature occured. When CoCrFeNi alloy is cold rolled, the magnetic entropy change for a change in applied field of 2T is $\Delta S_m = -0.35$ J/kg K. Cold rolling results in a broadening of ΔS_m vs. temperature curve. A heat treatment after cold rolling at 1073 K sharpens the magnetic entropy curve. It is found that upon heating (after cold rolling) there is a reentrant magnetic moment near 730 K. This feature is much less pronounced in the as-cast samples (without cold rolling) and in the Cr-rich samples, and is no longer observed after annealing at 1073 K. This reentrant magnetic moment upon heating is argued to be due to changes in short range order or the coalescence of vacancies forming stacking faults (Lucas et al., 2013).

9.6 HYDROGEN STORAGE PROPERTIES

Hydrogen is widely considered a renewable and sustainable solution for reducing the consumption of fossil fuels. Metal hydrides have received particular attention for their excellent hydrogen storage properties. Several HEA systems have been investigated for their hydrogen storage properties.

Kao et al. (2010) have used pressure-composition-isotherms (PCIs) to demonstrate that $CoFeMnTi_xVZr$, $CoFeMnTiV_yZr$, and $CoFeMnTiVZr_z$ can absorb and desorb hydrogen for compositions of $0.5 \leq x \leq 2.5$, $0.4 \leq y \leq 3.0$, and $0.4 \leq z \leq 3.0$. Because of the high entropy effect in the alloy system, each HEA is a multicomponent solid solution with a single C14 Laves phase structure before and after PCI test. The lattice expands when hydrides form. Maximum hydrogen storage capacity $(H/M)_{max}$ could be up to 1.8 wt% at room temperature for $CoFeMnTi_2VZr$ alloy. This study also elucidated the effects of the chemical composition on the properties of hydrogen storage in terms of lattice constant, element segregation, and hydride formation enthalpies. The factor that determines the maximum hydrogen storage capacity is the affinity between the alloying elements and hydrogen. In $CoFeMnTi_xVZr$ and $CoFeMnTiVZr_z$, Ti and Zr additions promote hydrogen absorption. The decrease in $(H/M)_{max}$ in alloys $Ti_{2.5}$ and $Zr_{3.0}$ can be attributed to the strong segregation of Ti and

Zr in the cast structure. Following hydrogen desorption, a large amount of hydrogen is retained in Zr_z alloys due to the strong bonding between Zr and H. By calculating the average formation enthalpy of hydride, they show that (H/atom)$_{max}$ is a linear function of ΔH_{cal} in A_2B and AB_5 alloy systems as shown in Figure 9.4. Although it is nonlinear with Ti_x, V_y, Zr_z, AB, and AB_2, (H/atom)$_{max}$ is strongly related to ΔH_{cal} for Ti_x, V_y, and Zr_z. From this study, it is observed that multicomponent HEAs with C14 Laves phase structure can be obtained by varying Ti, V, and Zr contents with good hydrogen storage potential.

Kunce et al. (2013) prepared ZrTiVCrFeNi HEA by LENS technique. Pressure-composition-temperature (PCT) isotherms were measured up to 100 bar of hydrogen at 50°C following activation of the alloy by annealing at 500°C for 2 h under vacuum, as shown in Figure 9.5. The maximum hydrogen adsorption capacity was 1.81 wt% after synthesis and 1.56 wt% after annealing. It is found that the equilibrium pressure of hydrogen desorption was too low for a complete desorption reaction, resulting in the presence of a C14 hydride phase in the alloy after PCT tests. This indicates that a thorough understanding of the structural and accompanying hydrogen storage property changes is of fundamental importance for future development of HEAs as potential materials for hydrogen storage (Kunce et al., 2013).

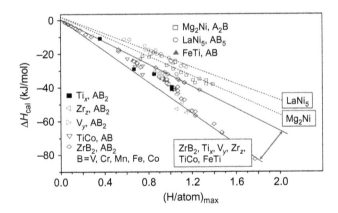

Figure 9.4 (Hlatom)$_{max}$ versus ΔH_{cal} curves for Mg_2Ni, $LaNi_5$, $TiFe$ (AB), $TiCo$ (AB), ZrB_2 (AB_2), Ti_x, V_y, and Zr_z. The curve shows that (Hlatom)$_{max}$ is a linear function of ΔH_{cal} for Mg_2Ni, $LaNi_5$, but nonlinear for AB, AB_2, Ti_x, V_y, and Zr_z (Kao et al., 2010).

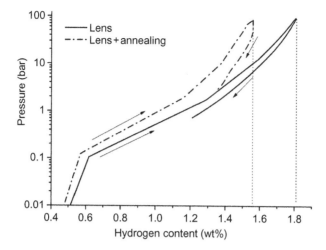

Figure 9.5 PCT absorption and desorption curves at 50°C for ZrTiVCrFeNi alloy following LENS deposition with additional annealing at 1000°C for 24 h (Kunce et al., 2013).

9.7 IRRADIATION RESISTANCE

It is known that high energy particle irradiation on solids produces atomic displacements and thermal spikes. The high atomic-level stresses in HEAs facilitate amorphization upon particle irradiation, followed by local melting and recrystallization due to thermal spikes. It is speculated that this process will leave much less defects in HEAs than in conventional alloys. For this reason, they may be excellent candidates for some nuclear applications. Initial results of computer simulation on model binary alloys and an electron microscopy study on Hf–Nb–Zr alloys, which demonstrate extremely high irradiation resistance of these alloys against electron damage to support this speculation, were reported by Egami et al. (2013). Simulations on Hf–Nb–Zr alloys lead to a conclusion that HEAs with the atomic-level volume strain close to 0.1 are self-healing and highly irradiation resistant.

9.8 CATALYTIC PROPERTIES

Nanoparticles of $Pt_xFe_{(100-x)/5}Co_{(100-x)/5}Ni_{(100-x)/5}Cu_{(100-x)/5}Ag_{(100-x)/5}$ ($x = 22, 29, 52, 56$) prepared by the sputter depositions on pretreated carbon clothes were investigated for their electrocatalytic ability for methanol oxidation (Tsai et al., 2009a). Cyclic voltammetry of the deposited samples demonstrated enhancements in the current response with increasing Pt content and deposition time. $Pt_{52}Fe_{11}Co_{10}Ni_{11}Cu_{10}Ag_8$ alloy exhibited the highest mass activity of $462-504$ mA/mg. Identical process was conducted to fabricate electrodes with sputtered Pt and $Pt_{43}Ru_{57}$. In comparison, the $Pt_{52}Fe_{11}Co_{10}Ni_{11}Cu_{10}Ag_8$ showed moderate improvements over that of Pt but was still outperformed by the $Pt_{43}Ru_{57}$.

CHAPTER 10

Applications and Future Directions

10.1 INTRODUCTION

From previous chapters, it is evident that HEA concept has opened up an entirely new alloy field, which encompasses a wide range of microstructure and properties. Not only their compositions could be designed but also conventional (ingot metallurgy, powder metallurgy, and coating technology) processes or even new processes could be selected to generate unique alloys with different properties. The new processes include thermomechanical treatments, rapid solidification, mechanical alloying, spray forming, semisolid forming, equal channel angular extrusion, reciprocating extrusion, superplastic forming, high-strain-rate superplastic forming, stir friction welding, spark plasma sintering, and nanoscale materials technology. As a result, numerous combinations and possibilities exist. From current HEA-related literature, it is clear that a suitable alloy composition and proper processing might obtain outstanding properties for intended applications. Besides HEAs research for scientific curiosity, researchers expect that HEAs can substitute conventional materials in difficult and stringent operating conditions by providing superior performance with increased service. This ultimately improves energy saving, materials saving, performance efficiency, cost reduction, resources conservation, environment, and health. This chapter includes the property goals pursued, advanced applications demanding new materials, and a few examples. This chapter also reviews the current state of patents related to HEAs and points out future directions of HEAs.

10.2 GOALS OF PROPERTY IMPROVEMENT

Materials selection involves seeking the best match between the property profiles of the materials in the universe and that required by design (Ashby, 2011). But from engineering and commercial points of view, materials properties and cost are the main consideration in selection of materials to fit the requirements. In other words, the main aim is to minimize cost while meeting product performance goals.

In addition, manufacturability, availability, appearance, and recyclability are also factors (Kalpakjian and Schmid, 2014) to be taken into account.

There are many properties which are investigated in materials research. It is often noted that some properties are emphasized and pursued in order to attain higher levels than the existing ones. For example (Yeh, 2006):

1. Strength and toughness: higher combination of strength and toughness
2. Wear resistance: higher adhesion and abrasion wear resistances, and higher erosion resistance
3. High-temperature resistance: higher softening, oxidation, hot corrosion (sulfidation), thermal shock, thermal fatigue, and creep resistances, and high microstructural stability
4. Chemical resistance: improved resistance to general corrosion, pitting corrosion, and stress corrosion
5. Radiation resistance: high radiation resistance for the structure in nuclear reactors
6. Light weight: low density for energy saving and high efficiency
7. Formability: good superplasticity and high-strain-rate superplasticity for materials saving and weight reduction
8. Magnetism: superior soft or hard magnetic properties and higher Curie points
9. Electrical resistance: low thermal coefficient of resistivity in a large temperature range
10. Thermal conductivity: high thermal conduction for heat spreaders but low thermal conduction for thermal barrier applications
11. Diffusion resistance: diffusion barriers to prevent interdiffusion and reaction between two materials
12. Green requirements: low-pollution, lead-free, cadmium-free, Cr^{+6}-free, recyclable, and low energy consumption

10.3 ADVANCED APPLICATIONS DEMANDING NEW MATERIALS

Although many advanced applications presently use conventional alloys, there always exists the demand of materials with higher and higher performance. There are various driving forces to push the development of new and improved materials including market pull and competition, curiosity-driven research, miniaturization, multifunctionality, environmental consciousness, and increased product liability (Ashby, 2011). For example, longer lifetime of components and systems means lower replacement cost and resource saving. In addition, higher energy conversion efficiency at higher temperature and higher pressure operation conditions in engines improves efficiency, which reduces fuel consumption, cost and pollution. In the current state, a number of advanced applications demanding new and improved materials include the following (Yeh, 2006):

1. Engine materials: higher elevated-temperature strength, oxidation resistance, hot corrosion resistance, and creep resistance
2. Nuclear materials: improved elevated-temperature strength and toughness with low irradiation damage
3. Tool materials and hard-facing materials: improved room and elevated-temperature strength and toughness, wear resistance, impact strength, low friction, corrosion resistance, and oxidation resistance
4. Waste incinerators: improved elevated-temperature strength, wear resistance, corrosion resistance, and oxidation resistance
5. Chemical plants: improved corrosion resistance, wear resistance and cavitation resistance for chemical piping systems, pumps, and mixers
6. Marine structures: improved corrosion resistance and erosion in seawater
7. Heat-resistant frames for multifloor buildings: higher elevated-temperature strength which could sustain during incidences of fire
8. Light transportation materials: improved specific strength and toughness, fatigue strength, creep resistance, and formability
9. High-frequency communication materials: high electrical resistance and magnetic permeability above 3 GHz
10. Functional coatings for 3C (Computers/Communications/Consumer, meaning electronic) products: better wear and corrosion resistances, antisticky, antifingerprint, antibacterial, and esthetics
11. Functional coatings for molds and tools: improved hot hardness, toughness, wear and corrosion resistances, and low friction coefficient
12. Hydrogen-storage materials for automobiles: low cost, high reversible volumetric and gravimetric density of hydrogen, and near-ambient cycling condition

13. Interconnect alloys for solid oxide fuel cell: high oxidation and creep resistances, low coefficient of thermal expansion, low area specific resistance, and good weldability
14. Superconductors: higher critical temperature and critical current
15. Thermoelectric materials: higher thermoelectric figure of merit at medium and high temperatures for converting waste heat into electricity
16. Electric and magnetic materials: constant thermal coefficient of resistivity and thermally stable magnetization in larger temperature range for precision electric and electronic devices
17. Golf-club-head materials: lower density, higher strength, and greater resilience

However, the development of conventional alloys and new alloys based on one or two major elements gradually approached their limit at the end of the twentieth century. This saturation has left difficulties in creating new materials to cater to the anticipated jump in performance. Under these circumstances, HEAs and related materials provide a new and huge opportunity. Most importantly, the reported outstanding properties of HEAs have shown that suitable composition design and process selection might yield HEAs replacing traditional materials for such applications. In fact, almost all the above items have been researched with HEAs, HEA-based ceramics, HEA-binder carbides, and HEA-binder cermets in recent years.

10.4 EXAMPLES OF APPLICATIONS

Several examples are introduced in this section to demonstrate the promising applications of HEAs.

10.4.1 $Al_5Cr_{12}Fe_{35}Mn_{28}Ni_{20}$ HEA

This alloy has a simple FCC structure with spherical NiAl-rich precipitates. In its cast state or homogenized state it displays a very good workability (Yeh, 2006). No intermediate annealing during cold rolling is required in contrast to the frequent annealing in 304 stainless steels. Figure 10.1 shows a foil of 70 μm thick which was cold rolled to an extension of 4257% without any edge cracking. Its work hardening curve indicates that its hardness approaches saturation around 360 HV at large deformation. At this state, the foils still can be folded without cracking. This indicates that its bendability is excellent. The alloy also

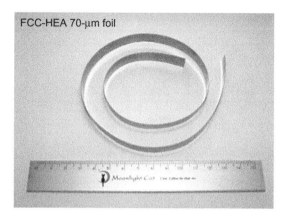

Figure 10.1 Seventy-μm-thick rolled foil of $Al_5Cr_{12}Fe_{35}Mn_{28}Ni_{20}$ alloy prepared by arc melting and cold rolling. The rolling extension is 4257% based on the work piece with an initial thickness of around 3 mm and a hardness of 147 HV.

shows good corrosion resistance in salt spray test. Insulating thin film of SiN_x of 800 nm in thickness was deposited on the polished surface of this foil for testing the minimum bend radius without causing permanent deformation of foil. Adhesion and current leakage of SiN_x thin film were also assessed. The minimum radius is 10 mm at which no spalling occurs. The leakage current after bending test is 8×10^{-9} A/cm^2. This easily rollable alloy thus has potential application as flexible substrates for solar cells and displays.

10.4.2 $Co_{1.5}CrFeNi_{1.5}Ti_{0.5}$ HEA

This alloy has a simple FCC structure in the as-cast state. It can be heat-treated to obtain gamma prime and eta precipitates in the FCC matrix. The as-cast hardness of the alloy is 378 HV and a peak hardness of 513 HV can be obtained by aging at 800°C for 5 h. It shows smaller magnetization and higher resistivity than Stellite 6. It has a very small eddy current loss under alternating magnetic field. It also shows better abrasion wear resistance tested by the pin-on-belt method with alumina sand belt and a higher corrosion resistance in 0.5 M $H_2SO_4 + 0.5$ M NaCl solution than Stellite 6. Figure 10.2 shows the cast and machined bearings made by lost-wax cast method. It can be used in severe conditions such as underground electrical pump components used in oil well system. Because of its high temperature strength and oxidation resistance, this alloy has been used for components such as connecting rods and holders for high-temperature tensile testing

Figure 10.2 Two machined cast bearings of $Co_{1.5}CrFeNi_{1.5}Ti_{0.5}$ HEA made by lost-wax cast method. The outer diameter is 56 mm and inner diameter is 38 mm.

machine up to 1000°C at which its hot hardness is still high, around 250 HV, which is higher than that of most commercial superalloys.

10.4.3 Profile Hardening of $Al_{0.3}CrFe_{1.5}MnNi_{0.5}$ HEA

$Al_{0.3}CrFe_{1.5}MnNi_{0.5}$ HEA has shown significant age hardening at 600–800°C. The hardening and thus improved wear resistance is due to the formation of ρ-phase which is a Cr-, Fe-, and Mn-rich phase. By exploiting the merit of high-temperature aging, the alloy parts could be profile hardened by simple heating. Figure 10.3 shows a comb-shaped part which was profile-hardened by simple heating in air at 550°C for 2 h (Chuang et al., 2013). All the surfaces get uniform hardened layer around 74 μm in thickness. The surface layer has a hardness of 1090 HV whereas the substrate has a hardness of 338 HV. The abrasion wear resistance is about 1.45 times that of SKD61 tool steel (520 HV) and 1.3 times that of SUJ2 bearing steel (723 HV). This demonstrates that the HEA is unique in providing an effective route for surface hardening instead of shot peening, carburizing and nitriding treatment. It could be used for shafts, and also complex components requiring high strength and wear resistance.

10.4.4 HEA Coatings for Antisticky Molds and Solar Cells

Because HEA coatings easily form amorphous structure with very low roughness, they can be used for antisticky coating and diffusion-barrier applications. Figure 10.4 compares the hard-Cr coating and HEA coating on the fine-sand-blasted surface of a SKD61 steel mold for IC (integrated circuit) package. It can be seen that hard-Cr coating is

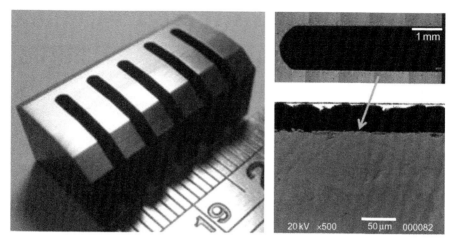

Figure 10.3 Profile-hardened comb-shaped component and the hardening surface layer along the inner surface of a notch (Chuang et al., 2013).

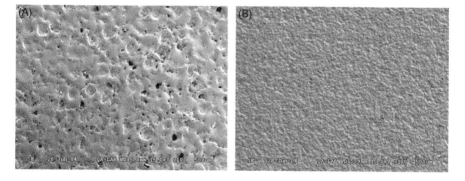

Figure 10.4 (A) Commercial hard-Cr coating and (B) AlCoCrCuFeNiTi HEA coating on fine-sand-blasted surface of SKD61 mold.

rougher than AlCoCrCuFeNiTi HEA coating. As a result, the release force of HEA coating required to separate molded IC components (packaged with the molding compound of epoxy resin, graphite, and silica particles) from the mold is only half that of the hard-Cr coating. This is also seen in the molding of fuel cell bipolar plates at 200°C; HEA coating is easy to release the molded plates as shown in Figure 10.5 whereas hard-Cr coating would cause warping and even tearing of molded plate. In addition, HEA film has been tested for CuInGaSe (CIGS) thin-film solar cell as back electrode in view of its thermally stable amorphous structure, better diffusion-barrier property, and higher reflectivity and conductivity as compared with Mo back

Figure 10.5 (A) Used SKD61 compression mold with HEA coating on the mold surface and (B) molded compos-ite bipolar plate (containing polymer and fillers) 1.5 mm in thickness without warping or tear.

electrode. The energy conversion efficiency is effectively increased by 9% based on the undisclosed data of a CIGS developing company (Yeh, 2013b).

10.4.5 HEA Solders for Welding Hard Metal and Steel

Because copper-based brazing alloy for welding cemented carbide and steel tends to fail due to lower strength or excessive corrosion, a HEA brazing filler, for welding cemented carbide and steel, having excellent strength, toughness and corrosion resistance, spreadability, and bonding strength has been developed (Zhai and Xu, 2011). The brazing foil with several tens of micrometers is produced by single-roller melt spinning in a vacuum chamber. The composition range is 10−15% Ti, 18−25% Cu, 12−18% Ni, 10−15% Zr, 15−20% Fe, 10−15% Cr, 0.5−2.5% Sn, and 0.01−2% of trace elements selected from Bi, Ga, or In. The strength of the weld is around 200 MPa. This invention largely improves the perfor-mance of conventional brazing fillers for cutting tools.

From these examples, it can be seen HEAs might have versatile applications and have improved performance in substituting con-ventional alloys. Like the progress of traditional alloys requiring long-term development and improvement, HEAs also need more investigation and research to develop and fine-tune their microstruc-ture and properties.

10.5 PATENTS ON HEAs AND RELATED MATERIALS

In the huge scope of HEAs and HEA-related materials, a number of patents have been applied and granted. The first patent with the title

"High-entropy multicomponent alloys" concerning the composition range of HEAs from five to eleven principal elements was filed as early as 1998 for a Taiwan patent which was granted in 2003 as Taiwan patent number 193729. It was also granted by China (patent number 00133500.6) and Japan (patent number 4190720) but with a smaller composition range. However, the patent was not granted by the United States. After this patent, there are more patents filed and granted. Huang (2009) reviewed the status of HEA patents in 2009. He divided the patents into two groups. One consists of HEAs having some specific functions, the other consists of composite materials with HEAs and other reinforcements. The first group includes hard-facing alloy, soft-magnetic thin film inductor and magnetic multicomponent alloy film, HEA catalyst, sprayed HEA coating, high-temperature HEAs with low concentration of cobalt and nickel. The second group includes the cemented carbide with HEA binders, and the HEA-based complex materials.

After 2009, there has been a large increase in patent applications. Among these, most are related to cemented carbides, cermets, and spray-deposited hard-facing materials. Like the patent CN100526490C (China patent publication number is CN1827817A) disclosed in 2006, three patent applications, CN102787266A, CN102787267A, and CN102796933A are related to hard metals and cermets with HEA binders. US patent 8075661 deals with ultrahard composite materials and their manufacturing method. China patent CN100535150C deals with hard-facing HEAs. CN102828139A concerns with HEA powders for spray coatings. CN103255414A, CN103255415A, and CN103276276A deal with carbide-reinforced HEA spray coatings deposited by plasma spray technique.

On the other hand, a composition range of hydrogen-storage HEAs has been granted in Taiwan I402357 and applied in China and the United States for a patent. Two composition ranges of HEA brazing fillers has been granted as China patents CN101554686B and CN101590574B, respectively. Refractory HEAs patent entitled with "multicomponent solid solution alloys having high mixing entropy" has been applied in the United States (US20130108502 A1). HEA piston ring patent entitled with "multicomponent alloy base piston ring" has been granted as Taiwan patent I403594. Based on this trend, it can be predicted that a large number of HEA-related material patents will be generated by more and more research and development in the future.

10.6 FUTURE DIRECTIONS

In the last decade, more than 500 HEA journal and conference papers have been published, Nevertheless the understanding of the whole HEA world is still in its infancy. Several future research trends can be foreseen (Yeh, 2013b).

More fundamental and basic studies are required. Because materials science and solid state physics are mainly based on conventional materials with one or two principal elements, what happens in HEAs would be interesting for better understanding of materials. In the whole-solute matrix, different contributions to mixing entropy, mutual interactions in all unlike atomic pairs, short range order, lattice distortion, electrical and thermal conductivity, thermal expansion, vacancy concentration, diffusion coefficients, phase transformation, Young's modulus, dislocation energy, staking fault energy, grain boundary energy, slip, twinning, serration behavior, strengthening, toughening, fracture, fatigue, creep, wear, corrosion, and oxidation are all needed to be understood with their mechanisms. Whether these mechanisms are simple extensions from those of conventional alloys or not is a matter of inquiry.

More research on composites of HEAs with ceramic reinforcements and high-entropy ceramic (HEC) reinforcements is required. Such a combination would generate numerous composites among which many opportunities could be found for critical applications not easily attained by traditional composites.

More research on medium-entropy alloys (MEAs) is also required. It is recognized that there still exists a large space in MEAs. This is reasonable since the view from the high-entropy side and that from conventional low-entropy side could be linked up to generate more sparks in the undeveloped area of MEAs which could be Fe-base, Ni-base, Co-base, Cr-base, and Cu-base. In addition to different routes of alloy design strategy, suitable compositions of MEAs might be easily obtained from those phases that are rich in some specific elements during the investigation of microstructure and its correlation with properties for HEAs. This new finding can be regarded as a by-product from HEAs research.

Assessment of existing database to find possible applications is required. It is believed that HEAs, HECs, or their composites could solve many bottlenecks encountered by conventional materials. Although the database is still limited, many suggestions of their potential applications are seen from the literature.

One foresees more synthesis based on combinatorial methods which allow high-throughput exploration of the composition space. In addition to the preparation of the alloys, this allows the measurement of a variety of properties across the composition spectrum. Optimum composition with required properties could be found more efficiently. Moreover, their structure and microstructure can be effectively investigated. For example, the lattice distortion arising from the simultaneous presence of multiple solutes needs to be measured. The effect of this distortion on dislocation movement needs to be understood. More research with modelling and simulations are required. Remarkable progress can be anticipated as new computational tools, integrated computational materials engineering (ICME) and materials genome approach are employed.

In conclusion, HEAs and HE-related materials have potential applications in different fields and are expected to replace traditional materials in many sectors. In just a decade from 2004, extraordinary progress has been made. This research theme has caught global attention. A bright future is seen.

REFERENCES

Alonso, J.A., Simozar, S., 1980. Prediction of solid solubility in alloys. Phys. Rev. B 22, 5582–5589.

Ashby, M.F., 2008. Materials: a brief history. Philos. Mag. Lett. 88, 749–755.

Ashby, M.F., 2011. Materials Selection in Mechanical Design, fourth ed. Butterworth-Heinemann, Elsevier, Oxford, UK.

ASM Handbook, 1992. Alloy Phase Diagrams, vol. 3. ASM International, Materials Park, OH, USA.

Asta, M., Beckermann, C., Karma, A., Kurz, W., Napolitano, R., Plapp, M., et al., 2009. Solidification microstructures and solid-state parallels: recent developments, future directions. Acta Mater. 57, 941–971.

Benjamin, J.S., 1970. Dispersion strengthened superalloys by mechanical alloying. Metall. Mater. Trans. B 1, 2943–2944.

Bhatt, J., Jiang, W., Junhai, X., Qing, W., Dong, C., Murty, B.S., 2007. Optimization of bulk metallic glass forming composition in Zr–Cu–Al system by thermodynamic modelling. Intermetallics 15, 716–721.

Bhattacharjee, P.P., Sathiaraj, G.D., Zaid, M., Gatti, J.R., Lee, C., Tsai, C.W., et al., 2014. Microstructure and texture evolution during annealing of equiatomic CoCrFeMnNi high-entropy alloy. J. Alloys Compd. 587, 544–552.

Biswas, T., Ranganathan, S., 2006. Multicomponent alloys visualized as lower order alloys: examples of quasicrystals and metallic glasses. Ann. Chim. Sci. Mat. 31, 649–656.

Boesch, W.L., Slaney, J.S., 1964. Preventing sigma phase embrittlement in nickel base superalloys. Met. Prog. 86, 109–111.

Boettinger, W.J., Warren, J.A., Beckermann, C., Karma, A., 2002. Phase-field simulation of solidification. Annu. Rev. Mater. Res. 32, 163–194.

Braic, V., Balaceanu, M., Braic, M., Vladescu, A., Panseri, S., Russo, A., 2012. Characterization of multi-principal-element (TiZrNbHfTa)N and (TiZrNbHfTa)C coatings for biomedical applications. J. Mech. Behav. Biomed. Mater. 10, 197–205.

Cahn, R.W., 2001. The Coming of Materials Science. Elsevier Science Ltd., Amsterdam.

Callister, W.D., 2003. Materials Science and Engineering: An Introduction. John Wiley & Sons, New York, USA.

Cantor, B., 2007. Stable and metastable multicomponent alloys. Ann. Chim. Sci. Mat. 32 245–256.

Cantor, B., Kim, K.B., Warren, P.J., 2002. Novel multicomponent amorphous alloys. Mater. Sci. Forum 386–388, 27–32.

Cantor, B., Chang, I.T.H., Knight, P., Vincent, A.J.B., 2004. Microstructural development in equiatomic multicomponent alloys. Mater. Sci. Eng. A 375–377, 213–218.

Chang, H.W., Huang, P.K., Davison, A., Yeh, J.W., Tsau, C.H., Yang, C.C., 2008. Nitride films deposited from an equimolar Al–Cr–Mo–Si–Ti alloy target by reactive direct current magnetron sputtering. Thin Solid Films 516, 6402–6408.

Chang, S.Y., Chen, M.K., Chen, D.S., 2009. Multiprincipal-element AlCrTaTiZr-nitride nano-composite film of extremely high thermal stability as diffusion barrier for Cu metallization. J. Electrochem. Soc. 156, G37−G42.

Chang, S.Y., Wang, C.Y., Chen, M.K., Li, C.E., 2011. Ru incorporation on marked enhance-ment of diffusion resistance of multi-component alloy barrier layers. J. Alloys Compd. 509, L85−L89.

Chelikowsky, J.R., 1979. Solid solubilities in divalent alloys. Phys. Rev. B 19, 686−701.

Chen, H.C., Jan, D.J., Chen, C.H., Huang, K.T., 2013a. Bond and electrochromic properties of WO$_3$ films deposited with horizontal DC, pulsed DC, and RF sputtering. Electrochim. Acta 93, 307−313.

Chen, H.S., 1974. Thermodynamic considerations on the formation and stability of metallic glasses. Acta Metall. 22, 1505−1511.

Chen, J.H., Chen, P.N., Hua, P.H., Chen, M.C., Chang, Y.Y., Wu, W., 2009a. Deposition of multicomponent alloys on low-carbon steel using gas tungsten arc welding (GTAW) cladding process. Mater. Trans. 50, 689−694.

Chen, J.H., Chen, P.N., Lin, C.M., Chang, C.M., Chang, Y.Y., Wu, W., 2009b. Microstructure and wear properties of multicomponent alloy cladding formed by gas tungsten arc welding (GTAW). Surf. Coat. Technol. 203, 3231−3234.

Chen, J.H., Hua, P.H., Chen, P.N., Chang, C.M., Chen, M.C., Wu, W., 2008. Characteristics of multi-element alloy cladding produced by TIG process. Mater. Lett. 62, 2490−2492.

Chen, K.Y., Shun, T.T., Yeh, J.W., 2002. Development of Multi-Element High-Entropy Alloys for Spray Coating, Master's thesis. National Tsing Hua University, Taiwan.

Chen, M.J., Lin, S.S., 2003. The Effect of V, S, and Ti Additions on the Microstructure and Wear Properties of Al$_{0.5}$CrCuFeCoNi High-Entropy Alloys, Master's thesis. National Tsing Hua University, Taiwan.

Chen, M.R., Lin, S.J., Yeh, J.W., Chen, S.K., Huang, Y.S., Chuang, M.H., 2006a. Effect of vanadium addition on the microstructure, hardness and wear resistance of Al$_{0.5}$CoCrCuFeNi high-entropy alloy. Metall. Mater. Trans. A 37, 1363−1369.

Chen, M.R., Lin, S.J., Yeh, J.W., Chen, S.K., Huang, Y.S., Tu, C.P., 2006b. Microstructure and properties of Al$_{0.5}$CoCrCuFeNiTi$_x$ ($x = 0-2.0$) high-entropy alloys. Mater. Trans. 47, 1395−1401.

Chen, S.K., Kao, Y.F., 2012. Near-Constant Resistivity in 4.2−360 K in a B2 Al$_{2.08}$CoCrFeNi. Am. Inst. Phys. Advances 2, 012111-1-5.

Chen, S.T., Tang, W.Y., Kuo, Y.F., Chen, S.Y., Tsau, C.H., Shun, T.T., et al., 2010a. Microstructure and properties of age-hardenable Al$_x$CrFe$_{1.5}$MnNi$_{0.5}$ alloys. Mater. Sci. Eng. A 527, 5818−5825.

Chen, T.K., Shun, T.T., Yeh, J.W., Wong, M.S., 2004. Nanostructured nitride films of multi-element high-entropy alloys by reactive DC sputtering. Surf. Coat. Technol. 188−189, 193−200.

Chen, T.K., Wong, M.S., Shun, T.T., Yeh, J.W., 2005a. Nanostructured nitride films of multi-element high-entropy alloys by reactive DC sputtering. Surf. Coat. Technol. 200, 1361−1365.

Chen, W., Fu, Z., Fang, S., Xiao, H., Zhu, D., 2013b. Alloying behaviour, microstructure and mechanical properties in a FeNiCrCo$_{0.3}$Al$_{0.7}$ high entropy alloy. Mater. Des. 51, 854−860.

Chen, Y.L., Hu, Y.H., Hsieh, C.A., Yeh, J.W., Chen, S.K., 2009c. Competition between ele-ments during mechanical alloying in an octonary multi-principal-element alloy system. J. Alloys Compd. 481, 768−775.

Chen, Y.L., Hu, Y.H., Tsai, C.W., Hsieh, C.A., Kao, S.W., Yeh, J.W., et al., 2009d. Alloying behaviour of binary to octonary alloys based on Cu−Ni−Al−Co−Cr−Fe−Ti−Mo during mechanical alloying. J. Alloys Compd. 477, 696−705.

Chen, Y.L., Hu, Y.H., Hsieh, C.A., Yeh, J.W., Chen, S.K., 2009e. Competition between elements during mechanical alloying in an octonary multi-principal-element alloy system. J. Alloys Compd. 481, 768–775.

Chen, Y.L., Tsai, C.W., Juan, C.C., Chuang, M.H., Yeh, J.W., Chin, T.S., et al., 2010b. Amorphization of equimolar alloys with HCP elements during mechanical alloying. J. Alloys Compd. 506, 210–215.

Chen, Y.Y., Duval, T., Hung, U.D., Yeh, J.W., Shih, H.C., 2005b. Microstructure and electrochemical properties of high entropy alloys-a comparison with type-304 stainless steel. Corros. Sci. 47, 2257–2279.

Chen, Y.Y., Hong, U.T., Shih, H.C., Yeh, J.W., Duval, T., 2005c. Electrochemical kinetics of the high entropy alloys in aqueous environments - A comparison with type 304 stainless steel. Corros. Sci. 47, 2679–2699.

Chen, Y.Y., Hong, U.T., Yeh, J.W., Shih, H.C., 2005d. Mechanical properties of a novel bulk $Cu_{0.5}NiAlCoCrFeSi$ glassy alloy in 288°C high-purity water. Appl. Phys. Lett. 87, 261918-1–261918-3.

Chen, Y.Y., Hong, U.T., Yeh, J.W., Shih, H.C., 2006c. Selected corrosion behaviour of a $Cu_{0.5}NiAlCoCrFeSi$ bulk glassy alloy in 288°C high-purity water. Scr. Mater. 54, 1997–2001.

Cheng, J.B., Liang, X.B., Wang, Z.H., Xu, B.S., 2013. Formation and mechanical properties of CoNiCuFeCr high-entropy alloys coatings prepared by plasma transferred arc cladding process. Plasma Chem. Plasma Process. 33, 979–992.

Cheng, J.B., Liang, X.B., Xu, B.S., 2014. Effect of Nb addition on the structure and mechanical behaviours of CoCrCuFeNi high-entropy alloy coatings. Surf. Coat. Technol. 240, 184–190.

Chou, H.P., Chang, Y.S., Chen, S.K., Yeh, J.W., 2009. Microstructure, thermophysical and electrical properties in $Al_xCoCrFeNi$ ($0 \leq x \leq 2$) high-entropy alloys. Mater. Sci. Eng. B 163, 184–189.

Chou, Y.L., Wang, Y.C., Yeh, J.W., Shih, H.C., 2010a. Pitting corrosion of the high-entropy alloy $Co_{1.5}CrFeNi_{1.5}Ti_{0.5}Mo_{0.1}$ in chloride-containing sulphate solutions. Corros. Sci. 52, 3481–3491.

Chou, Y.L., Yeh, J.W., Shih, H.C., 2010b. The effect of molybdenum on the corrosion behaviour of the high-entropy alloys $Co_{1.5}CrFeNi_{1.5}Ti_{0.5}Mo_x$ in aqueous environments. Corros. Sci. 52, 2571–2581.

Chou, Y.L., Yeh, J.W., Shih, H.C., 2011. Effect of inhibitors on the critical pitting temperature of the high-entropy alloy $Co_{1.5}CrFeNi_{1.5}Ti_{0.5}Mo_{0.1}$. J. Electrochem. Soc. 158, C246–C251.

Chuang, M.H., Tsai, M.H., Wang, W.R., Lin, S.J., Yeh, J.W., 2011. Microstructure and wear behaviour of $Al_xCo_{1.5}CrFeNi_{1.5}Ti_y$ high-entropy alloys. Acta Mater. 59, 6308–6317.

Chuang, M.H., Tsai, M.H., Tsai, C.W., Yang, N.H., Chang, S.Y., Yeh, J.W., et al., 2013. Intrinsic surface hardening and precipitation kinetics of $Al_{0.3}CrFe_{1.5}MnNi_{0.5}$ multi-component alloy. J. Alloys Compd. 551, 12–18.

Cui, H., Wang, H., Wang, J., Fu, H., 2011a. Microstructure and micro segregation in directionally solidified FeCoNiCrAl high entropy alloy. Adv. Mater. Res. 189–193, 3840–3843.

Cui, H., Zheng, L., Wang, J., 2011b. Microstructure evolution and corrosion behaviour of directionally solidified FeCoNiCrCu high entropy alloy. Appl. Mech. Mater. 66–68, 146–149.

Cullity, B.D., Stock, S.R., 2001. Elements of X-Ray Diffraction. Prentice Hall, New Jersey, USA.

Curtarolo, S., Gus, L., Hart, W., Nardelli, M.B., Mingo, N., Sanvito, S., et al., 2013. The high-throughput highway to computational materials design. Nat. Mater. 12, 191–201.

Daoud, H.M., Manzoni, A., Völkl, R., Wanderka, N., Glatzel, U., 2013. Microstructure and tensile behaviour of $Al_8Co_{17}Cr_{17}Cu_8Fe_{17}Ni_{33}$ (at.%) high-entropy alloy. JOM 65, 1805–1814.

Darken, L.S., Gurry, R.W., 1953. Physical Chemistry of Metals. McGraw-Hill, New York, NY.

de Boer, F.R., Boom, R., Mattens, W.C.M., Miedema, A.R., Niessen, A.K., 1988. Cohesion in Metals: Transition Metal Alloys (Cohesion and Structure). North Holland, North Holland Physics Publishing. Amsterdam, The Netherlands.

de Graef, M.D., McHenry, M.E., 2012. Structure of Materials: An introduction to Crystallography, Diffraction and Symmetry. Cambridge University Press, Cambridge, UK.

Del Grosso, M.F., Bozzolo, G., Mosca, H.O., 2012. Determination of the transition to the high entropy regime for alloys of refractory elements. J. Alloys Compd. 534, 25−31.

Ding, H.Y., Yao, K.F., 2013. High entropy $Ti_{20}Zr_{20}Cu_{20}Ni_{20}Be_{20}$ bulk metallic glass. J. Non-Cryst. Solids 364, 9−12.

Dolique, V., Thomann, A.L., Brault, P., Tessier, Y., Gillon, P., 2009. Complex structure/composition relationship in thin films of AlCoCrCuFeNi high entropy alloy. Mater. Chem. Phys. 117, 142−147.

Dolique, V., Thomann, A.L., Brault, P., Tessier, Y., Gillon, P., 2010. Thermal stability of AlCoCrCuFeNi high entropy alloy thin films studied by in-situ XRD analysis. Surf. Coat. Technol. 204, 1989−1992.

Dong, Y., Lu, Y., Kong, J., Zhang, J., Li, T., 2013a. Microstructure and mechanical properties of multi-component $AlCrFeNiMo_x$ high-entropy alloys. J. Alloys Compd. 573, 96−101.

Dong, Y., Lu, Y., Zhang, J., Li, T., 2013b. Microstructure and properties of multi-component $Al_xCoCrFeNiTi_{0.5}$ high-entropy alloys. Mater. Sci. Forum 745−746, 775−780.

Drosback, M., 2014. Materials genome initiative: advances and initiatives. JOM 66, 334−335.

Durga, A., Hari Kumar, K.C., Murty, B.S., 2012. Phase formation in equiatomic high entropy alloys: CALPHAD approach and experimental studies. Trans. Indian Inst. Met. 65, 375−380.

Egami, T., Guo, W., Rack, P.D., Nagase, T., 2013. Irradiation resistance of multicomponent alloys. Metall. Mater. Trans. A 45, 180−183.

Egami, T.Y., Waseda, Y., 1984. Atomic size effect on the formability of metallic glasses. J. Non-Cryst. Solids 64, 113−134.

Fang, S., Chen, W., Fu, Z., 2014. Microstructure and mechanical properties of twinned $Al_{0.5}CrFeNiCo_{0.3}C_{0.2}$ high entropy alloy processed by mechanical alloying and spark plasma sintering. Mater. Des. 54, 973−979.

Fang, S.S., Lin, G.W., Zhang, J.L., Zhou, Z.Q., 2002. The maximum solid solubility of the transition metals in palladium. Int. J. Hydrogen Energy 27, 329−332.

Fazakas, É., Varga, B., Varga, L.K., 2013. Processing and properties of nanocrystalline CoCrFeNiCuAlTiXVMo (X = Zn, Mn) high entropy alloys by mechanical alloying. ISRN Mech. Eng.

Fecht, H.J., Han, G., Fu, Z., Johnson, W.L., 1990. Metastable phase formation in the Zr−Al binary system induced by mechanical alloying. J. Appl. Phys. 67, 1744−1748.

Feng, X., Tang, G., Gu, L., Ma, X., Sun, M., Wang, L., 2012. Preparation and characterization of TaNbTiW multi-element alloy films. Appl. Surf. Sci. 261, 447−453.

Feng, X., Tang, G., Ma, X., Sun, M., Wang, L., 2013. Characteristics of multi-element (ZrTaNbTiW)N films prepared by magnetron sputtering and plasma based ion implantation. Nucl. Instrum. Methods Phys. Res., Sect. B 301, 29−35.

Fix, G.J., Fasano, A., Primicerio, M., 1983. In: Fasano, A., Primicerio, M. (Eds.), Free Boundary Problems: Theory and Applications. Pitman, Boston, MA.

Fu, Z., Chen, W., Fang, S., Zhang, D., Xiao, H., Zhu, D., 2013a. Alloying behaviour and deformation twinning in a $CoNiFeCrAl_{0.6}Ti_{0.4}$ high entropy alloy processed by spark plasma sintering. J. Alloys Compd. 553, 316−323.

Fu, Z., Chen, W., Xiao, H., Zhou, L., Zhu, D., Yang, S., 2013b. Fabrication and properties of nanocrystalline $Co_{0.5}FeNiCrTi_{0.5}$ high entropy alloy by MA-SPS technique. Mater. Des. 44, 535–539.

Fultz, B., 2010. Vibrational thermodynamics of materials. Prog. Mater Sci. 55, 247–352.

Gali, A., George, E.P., 2013. Tensile properties of high- and medium-entropy alloys. Intermetallics 39, 74–78.

Gao, M.C., Alman, D.E., 2013. Searching for next single-phase high-entropy alloy compositions. Entropy 15, 4504–4519.

Gao, X.Q., Zhao, H.B., Ding, D.W., Wang, W.H., Bai, H.Y., 2011. High mixing entropy bulk metallic glasses. J. Non-Cryst. Solids 357, 3557–3560.

Gaskell, D.R., 1995. Introduction to the Thermodynamics of Materials. Taylor and Francis, London, UK.

Greer, A.L., 1993. Confusion by design. Nature 336, 303–304.

Gschneidner Jr., K.A., 1980. L. S (Larry) Darken's contributions to the theory of alloy formation and where we are today. In: Bennett, L.H. (Ed.), Theory of Alloy Phase Formation. The Metallurgical Society of AIME, Warrendale, pp. 1–39.

Guo, S., Liu, C.T., 2011. Phase stability in high entropy alloys: formation of solid-solution phase or amorphous phase. Prog. Nat. Sci.: Mater. Int. 21, 433–446.

Guo, S., Hu, Q., Ng, C., Liu, C.T., 2013a. More than entropy in high-entropy alloys: forming solid solutions or amorphous phase. Intermetallics 41, 96–103.

Guo, S., Ng, C., Lu, J., Liu, C.T., 2011. Effect of valence electron concentration on stability of FCC or BCC phase in high entropy alloys. J. Appl. Phys. 109, 103505-1–103505-5.

Guo, S., Ng, C., Liu, C.T., 2013b. Anomalous solidification microstructures in Co-free $Al_xCrCuFeNi_2$ high-entropy alloys. J. Alloys Compd. 557, 77–81.

Guo, S., Ng, C., Wang, Z., Liu, C.T., 2014. Solid solutioning in equiatomic alloys: limit set by topological instability. J. Alloys Compd. 583, 410–413.

Guo, W., Dmowski, W., Noh, J.Y., Rack, P., Liaw, P.K., Egami, T., 2013c. Local atomic structure of a high-entropy alloy: an X-Ray and neutron scattering study. Metall. Mater. Trans. A 44, 1994–1997.

Handbook Committee, 1990. Metals Handbook, tenth ed, vols. 1 and 2. ASM International, Metals Park, OH.

He, J.Y., Liu, W.H., Wang, H., Wu, Y., Liu, X.J., Nieh, T.G., et al., 2014. Effects of Al addition on structural evolution and tensile properties of the FeCoNiCrMn high-entropy alloy system. Acta Mater. 62, 105–113.

Hemphill, M.A., Yuan, T., Wang, G.Y., Yeh, J.W., Tsai, C.W., Chuang, A., et al., 2012. Fatigue behaviour of $Al_{0.5}CoCrCuFeNi$ high entropy alloys. Acta Mater. 60, 5723–5734.

Hongfei, S., Nana, G., Canming, W., Zhongli, L., Haiyun, Z., 2011. Study of the microstructure of high-entropy alloys $AlFeCuCoNiCrTi_x$ ($x = 0$, 0.5, 1.0). Appl. Mech. Mater. 66–68, 66–68.

Hsieh, K.C., Yu, C.F., Hsieh, W.T., Chiang, W.R., Ku, J.S., Lai, J.H., et al., 2009. The microstructure and phase equilibrium of new high performance high-entropy alloys. J. Alloys Compd. 483, 209–212.

Hsu, C.Y., Juan, C.C., Wang, W.R., Sheu, T.S., Yeh, J.W., Chen, S.K., 2011. On the superior hot hardness and softening resistance of $AlCoCr_xFeMo_{0.5}Ni$ high-entropy alloys. Mater. Sci. Eng. A 528, 3581–3588.

Hsu, C.Y., Juan, C.C., Chen, S.T., Sheu, T.S., Yeh, J.W., Chen, S.K., 2013a. Phase diagrams of high-entropy alloy system Al–Co–Cr–Fe–Mo–Ni. J. Met. 65, 1829–1839.

Hsu, C.Y., Juan, C.C., Sheu, T.S., Chen, S.K., Yeh, J.W., 2013b. Effect of aluminum content on microstructure and mechanical properties of $Al_xCoCrFeMo_{0.5}Ni$ high-entropy alloys. J. Met. 65, 1840−1847.

Hsu, C.Y., Sheu, T.S., Yeh, J.W., Chen, S.K., 2010a. Effect of iron content on wear behaviour of $AlCoCrFe_xMo_{0.5}Ni$ high-entropy alloys. Wear 268, 653−659.

Hsu, C.Y., Shun, T.T., Chen, S.W., Yeh, J.W., 2003. Alloying effect of Boron on the microstructure and high-temperature properties of $CuCoNiCrAl_{0.5}Fe$ alloys, Master's thesis. National Tsing Hua University, Taiwan.

Hsu, C.Y., Wang, W.R., Tang, W.Y., Chen, S.K., Yeh, J.W., 2010b. Microstructure and mechanical properties of new $AlCo_xCrFeMo_{0.5}Ni$ high-entropy alloys. Adv. Eng. Mater. 12, 44−49.

Hsu, C.Y., Yeh, J.W., Chen, S.K., Shun, T.T., 2004. Wear resistance and high-temperature compression strength of FCC $CuCoNiCrAl_{0.5}Fe$ alloy with boron addition. Metall. Mater. Trans. 35A, 1465−1469.

Hsu, U.S., Hung, U.D., Yeh, J.W., Chen, S.K., Huang, Y.S., Yang, C.C., 2007. Alloying behaviour of iron, gold and silver in AlCoCrCuNi-based equimolar high-entropy alloys. Mater. Sci. Eng. A 460−461, 403−408.

Hsu, Y.H., Chen, S.W., Yeh, J.W., 2000. A Study on the Multicomponent Alloy Systems with Equal-Mole FCC or BCC Elements, Master's thesis. National Tsing Hua University, Taiwan.

Hsu, Y.J., Chiang, W.C., Wu, J.K., 2005. Corrosion behaviour of $FeCoNiCrCu_x$ high-entropy alloys in 3.5% sodium chloride solution. Mater. Chem. Phys. 92, 112−117.

Hsueh, H.T., Shen, W.J., Tsai, M.H., Yeh, J.W., 2012. Effect of nitrogen content and substrate bias on mechanical and corrosion properties of high-entropy films $(AlCrSiTiZr)_{100-x}N_x$. Surf. Coat. Technol. 206, 4106−4112.

Huang, C., Zhang, Y., Vilar, R., 2011. Microstructure characterization of laser clad TiVCrAlSi high entropy alloy coating on Ti-6Al-4V substrate. Adv. Mater. Res. 154−155, 621−625.

Huang, C., Zhang, Y., Vilar, R., Shen, J., 2012. Dry sliding wear behaviour of laser clad TiVCrAlSi high entropy alloy coatings on Ti-6Al-4V substrate. Mater. Des. 41, 338−343.

Huang, K.H., Yeh, J.W., 1996. A Study on the Multicomponent Alloy Systems Containing Equal-Mole Elements Master's thesis. National Tsing Hua University, Taiwan.

Huang, P.K., Yeh, J.W., 2003. Research of Multi-Component High-Entropy Alloys for Thermal Spray Coating, Master's thesis. National Tsing Hua University, Taiwan.

Huang, P.K., Yeh, J.W., 2009. Effects of substrate bias on structure and mechanical properties of (AlCrNbSiTiV)N coatings. J. Phys. D: Appl. Phys. 42, 1−7.

Huang, P.K., Yeh, J.W., 2010. Inhibition of grain coarsening up to 1000°C in (AlCrNbSiTiV)N superhard coatings. Scr. Mater. 62, 105−108.

Huang, P.K., Yeh, J.W., Shun, T.T., Chen, S.K., 2004. Multi-principal-element alloys with improved oxidation and wear resistance for thermal spray coating. Adv. Eng. Mater. 6, 74−78.

Huang, Y.S., 2009. Recent patents on high-entropy alloy. Recent Pat. Mater. Sci. 2, 154−157.

Huang, Y.S., Chen, L., Lui, H.W., Cai, M.H., Yeh, J.W., 2007. Microstructure, hardness, resistivity and thermal stability of sputtered oxide films of $AlCoCrCu_{0.5}NiFe$ high-entropy alloy. Mater. Sci. Eng. A 457, 77−83.

Huang, Y.T., Chen, S.K., Yeh, J.W., 2001. A Study on the Cu-Ni-Al-Co-Cr-Fe-Si-Ti Multicomponent Alloy System, Master's thesis. National Tsing Hua University.

Hume-Rothery, W., 1967. Factors affecting the stability of metallic phases. In: Rudman, P.S., Stringer, J., Jaffee, R.I. (Eds.), Phase Stability in Metals and Alloys. McGraw-Hill, New York, NY.

Integrated Computational Materials Engineering (ICME), 2008. A Transformational Discipline for Improved Competitiveness and National Security. National Research Council, The National Academies Press, Washington DC, USA.

Inoue, A., 1995. High strength bulk amorphous alloys with low critical cooling rates. Mater. Trans. JIM 36, 866−875.

Inoue, A., 1996. Recent progress of Zr-based bulk amorphous alloys, Science Reports of the Research Institutes, Tohoku University- Series A, vol. 42, pp. 1−11.

Inoue, A., Zhang, T., Masumoto, T., 1989. Al−La−Ni amorphous alloys with a wide supercooled region. Mater. Trans. JIM 30, 965−972.

International Technology Roadmap for Semiconductors, 2009.

Jeevan, H.S., Ranganathan, S., 2004. A new basis for the classification of quasicrystals. J. Non-Cryst. Solids 334−335, 184−189.

Ji, W., Fu, Z., Wang, W., Wang, H., Zhang, J., Wang, Y., et al., 2014. Mechanical alloying synthesis and spark plasma sintering consolidation of CoCrFeNiAl high-entropy alloy. J. Alloys Compd. 589, 61−66.

Juan, C.C., Hsu, C.Y., Tsai, C.W., Wang, W.R., Sheu, T.S., Yeh, J.W., et al., 2013. On microstructure and mechanical performance of AlCoCrFeMo$_{0.5}$Ni$_x$ high-entropy alloys. Intermetallics 32, 401−407.

Kalpakjian, S., Schmid, S., 2014. Manufacturing Engineering and Technology. Prentice Hall, New Jersey, USA.

Kao, S.W., Chen, Y.L., Chin, T.S., Yeh, J.W., 2006. A preliminary molecular dynamics simulation on equal-mole alloys with up to six elements. Ann. Chim. Sci. Mat. 31, 657−668.

Kao, S.W., Yeh, J.W., Chin, T.S., 2008. Rapidly solidified structure of alloys with up to eight equal-molar elements - a simulation by molecular dynamics. J. Phys. Condens. Matter. 20, 145214-1−145214-7.

Kao, Y.F., Chen, T.J., Chen, S.K., Yeh, J.W., 2009. Microstructure and mechanical property of as-cast, -homogenized and -deformed Al$_x$CoCrFeNi ($0 \leq \times \leq 2$) high-entropy alloys. J. Alloys Compd. 488, 57−64.

Kao, Y.F., Chen, S.K., Sheu, J.H., Lin, J.T., Lin, W.E., Yeh, J.W., et al., 2010. Hydrogen storage properties of multi-principal-component CoFeMnTi$_x$V$_y$Zr$_z$ alloys. Int. J. Hydrogen Energy 35, 9046−9059.

Kao, Y.F., Chen, S.K., Chen, T.J., Chu, P.C., Yeh, J.W., Lin, S.J., 2011. Electrical, magnetic and Hall properties of Al$_x$CoCrFeNi high-entropy alloys. J. Alloys Compd. 509, 1607−1614.

Kaufman, L., Agren, J., 2014. CALPHAD, first and second generation − Birth of the materials genome. Scr. Mater. 70, 3−6.

Kaufman, L., Cohen, M., 1956. The martensitic transformation in the Fe-Ni system. Trans. AIME 206, 1393−1401.

Kim, K.B., 2005. Formation of in-situ nanoscale Ag particles in (Ti$_{0.33}$Zr$_{0.33}$Hf$_{0.33}$)$_{40}$(Ni$_{0.33}$Cu$_{0.33}$Ag$_{0.33}$)$_{50}$Al$_{10}$ alloy with wide supercooled liquid region. Mater. Lett. 59, 1117−1120.

Kim, K.B., Warren, P.J., Cantor, B., 2003a. Formation of Metallic Glasses in Novel (Ti$_{33}$Zr$_{33}$Hf$_{33}$)$_{100-y}$(Ni$_{50}$Cu$_{50}$)$_x$Al$_y$ Alloys. Mater. Trans. 44, 411−413.

Kim, K.B., Warren, P.J., Cantor, B., 2003b. Metallic glass formation in multicomponent (Ti, Zr, Hf, Nb)−(Ni, Cu, Ag)−Al alloys. J. Non-Cryst. Solids 317, 17−22.

Kim, K.B., Zhang, Y., Warren, P.J., Cantor, B., 2003c. Crystallization behaviour in a new multicomponent Ti$_{16.6}$Zr$_{16.6}$Hf$_{16.6}$Ni$_{20}$Cu$_{20}$Al$_{10}$ metallic glass developed by the equiatomic substitution technique. Philos. Mag. 83, 2371−2381.

Kim, K.B., Warren, P.J., Cantor, B., 2004. Glass-forming ability of novel multicomponent $(Ti_{33}Zr_{33}Hf_{33})-(Ni_{50}Cu_{50})-Al$ alloys developed by equiatomic substitution. Mater. Sci. Eng. A 375-377, 317-321.

Kim, K.B., Warren, P.J., Cantor, B., Eckert, J., 2006a. Enhanced thermal stability of the devitrified nanoscale icosahedral phase in novel multicomponent amorphous alloys. J. Mater. Res. 21, 823-831.

Kim, K.B., Warren, P.J., Cantor, B., Eckert, J., 2006b. Structural evolution of nano-scale icosahedral phase in novel multicomponent amorphous alloys. Philos. Mag. 86, 281-286.

Kim, K.B., Yi, S., Hwang, I.S., Eckert, J., 2006c. Effect of cooling rate on microstructure and glass-forming ability of a $(Ti_{33}Zr_{33}Hf_{33})_{70}(Ni_{50}Cu_{50})_{20}Al_{10}$ alloy. Intermetallics 14, 972-977.

Kim, K.B., Warren, P.J., Cantor, B., Eckert, J., 2007. Devitrification of nano-scale icosahedral phase in multicomponent alloys. Mater. Sci. Eng. A 448-451, 983-986.

Klement, W., Willens, R.H., Duwez, P., 1960. Non-crystalline structure in solidified gold-silicon alloys. Nature 187, 869-870.

Koundinya, N.T.B.N., Sajith Babu, C., Sivaprasad, K., Susila, P., Kishore Babu, N., Baburao, J., 2013. Phase evolution and thermal analysis of nanocrystalline AlCrCuFeNiZn high entropy alloy produced by mechanical alloying. J. Mater. Eng. Perform. 22, 3077-3084.

Kunce, I., Polanski, M., Bystrzycki, J., 2013. Structure and hydrogen storage properties of a high entropy ZrTiVCrFeNi alloy synthesized using Laser Engineered Net Shaping (LENS). Int. J. Hydrogen Energy 38, 12180-12189.

Kuznetsov, A.V., Shaysultanov, D.G., Stepanov, N.D., Salishchev, G.A., Senkov, O.N., 2012. Tensile properties of an AlCrCuNiFeCo high-entropy alloy in as-cast and wrought conditions. Mater. Sci. Eng. A 533, 107-118.

Kuznetsov, A.V., Shaysultanov, D.G., Stepanov, N.D., Salishchev, G.A., Senkov, O.N., 2013. Superplasticity of AlCoCrCuFeNi high entropy alloy. Mater. Sci. Forum 735, 146-151.

Lai, C.H., Lin, S.J., Yeh, J.W., Chang, S.Y., 2006a. Preparation and characterization of AlCrTaTiZr multi-element nitride coatings. Surf. Coat. Technol. 201, 3275-3280.

Lai, C.H., Lin, S.J., Yeh, J.W., Davison, A., 2006b. Effect of substrate bias on the structure and properties of multi-element (AlCrTaTiZr)N coatings. J. Phys. D: Appl. Phys. 39, 4628-4633.

Lai, K.T., Chen, S.W., Yeh, J.W., 1998. Properties of the Multicomponent Alloy System with Equal-Mole Elements, Master's thesis. National Tsing Hua University, Taiwan.

Langer, J.S., Grinstein, G., Mazenko, G., 1986. Models of pattern formation in first-order phase transitions. In: Grinstein, G., Mazenko, G. (Eds.), Directions in Condensed Matter Physics. World Scientific, Singapore.

Lee, C.F., Shun, T.T., 2013. Age hardening of the $Al_{0.5}CoCrNiTi_{0.5}$ high-entropy alloy. Metall. Mater. Trans. A 45, 191-195.

Lee, C.P., Chen, Y.Y., Hsu, C.Y., Yeh, J.W., Shih, H.C., 2007. The effect of boron on the corrosion resistance of the high entropy alloys $Al_{0.5}CoCrCuFeNiB_x$. J. Electrochem. Soc. 154, C424-C430.

Lee, C.P., Chang, C.C., Chen, Y.Y., Yeh, J.W., Shih, H.C., 2008a. Effect of the aluminium content of $Al_xCrFe_{1.5}MnNi_{0.5}$ high-entropy alloys on the corrosion behaviour in aqueous environments. Corros. Sci. 50, 2053-2060.

Lee, C.P., Chen, Y.Y., Hsu, C.Y., Yeh, J.W., Shih, H.C., 2008b. Enhancing pitting corrosion resistance of $Al_xCrFe_{1.5}MnNi_{0.5}$ high-entropy alloys by anodic treatment in sulfuric acid. Thin Solid Films 517, 1301-1305.

Li, B.S., Wang, Y.P., Ren, M.X., Yang, C., Fu, H.Z., 2008a. Effects of Mn, Ti and V on the microstructure and properties of AlCrFeCoNiCu high entropy alloy. Mater. Sci. Eng. A 498, 482-486.

Li, B.Y., Peng, K., Hu, A.P., Zhou, L.P., Zhu, J.J., Li, D.Y., 2013a. Structure and properties of FeCoNiCrCu$_{0.5}$Al$_x$ high-entropy alloy. Trans Nonferrous Met. Soc. China 23, 735−741 (English Edition).

Li, C., Zhao, M., Li, J.C., Jiang, Q., 2008b. B2 structure of high-entropy alloys with addition of Al. J. Appl. Phys. 104, 113504−113506.

Li, C., Li, J.C., Zhao, M., Jiang, Q., 2009. Effect of alloying elements on microstructure and properties of multiprincipal elements high-entropy alloys. J. Alloys Compd. 475, 752−757.

Li, C., Li, J.C., Zhao, M., Jiang, Q., 2010a. Effect of aluminum contents on microstructure and properties of Al$_x$CoCrFeNi alloys. J. Alloys Compd. 504, S515−S518.

Li, H.F., Xie, X.H., Zhao, K., Wang, Y.B., Zheng, Y.F., Wang, W.H., et al., 2013b. In vitro and in vivo studies on biodegradable CaMgZnSrYb high-entropy bulk metallic glass. Acta Biomater. 9, 8561−8573.

Li, Q.H., Yue, T.M., Guo, Z.N., Lin, X., 2013c. Microstructure and corrosion properties of AlCoCrFeNi high entropy alloy coatings deposited on AISI 1045 steel by the electrospark process. Metall. Mater. Trans. A 44, 1767−1778.

Li, R., Gao, J., Fa, K., 2010b. Study to microstructure and mechanical properties of Mg containing high entropy alloys. Energy Environ Mater., Mater. Sci. Forum 650, 265−271.

Li, R., Gao, J.C., Fan, K., 2011. Microstructure and mechanical properties of MgMnAlZnCu high entropy alloy cooling in three conditions. Mater. Sci. Forum 686, 235−241.

Lin, C.H., Duh, J.G., Yeh, J.W., 2007. Multi-component nitride coatings derived from Ti−Al−Cr−Si−V target in RF magnetron sputter. Surf. Coat. Technol. 201, 6304−6308.

Lin, C.M., Tsai, H.L., 2011. Evolution of microstructure, hardness and corrosion properties of high-entropy Al$_{0.5}$CoCrFeNi alloy. Intermetallics 19, 288−294.

Lin, C.M., Tsai, H.L., Bor, H.Y., 2010. Effect of aging treatment on microstructure and properties of high-entropy Cu$_{0.5}$CoCrFeNi alloy. Intermetallics 18, 1244−1250.

Lin, P.C., Chin, T.S., Yeh, J.W., 2003. Development on the High Frequency Soft-Magnetic Thin Films from High-Entropy Alloys, Master's thesis. National Tsing Hua University, Taiwan.

Lin, Y.C., Cho, Y.H., 2008. Elucidating the microstructure and wear behaviour for multicomponent alloy clad layers by in situ synthesis. Surf. Coat. Technol. 202, 4666−4672.

Lin, Y.C., Cho, Y.H., 2009. Elucidating the microstructural and tribological characteristics of NiCrAlCoCu and NiCrAlCoMo multicomponent alloy clad layers synthesized in situ. Surf. Coat. Technol. 203, 1694−1701.

Liu, C.M., Wang, H.M., Zhang, S.Q., Tang, H.B., Zhang, A.L., 2014. Microstructure and oxidation behaviour of new refractory high entropy alloys. J. Alloys Compd. 583, 162−169.

Liu, L., Zhu, J.B., Zhang, C., Li, J.C., Jiang, Q., 2012. Microstructure and the properties of FeCoCuNiSn$_x$ high entropy alloys. Mater. Sci. Eng. A 548, 64−68.

Liu, W.H., Wu, Y., He, J.Y., Nieh, T.G., Lu, Z.P., 2013. Grain growth and the Hall-Petch relationship in a high-entropy FeCrNiCoMn alloy. Scr. Mater. 68, 526−529.

Liu, Z., Guo, S., Liu, X., Ye, J., Yang, Y., Wang, X.L., et al., 2011. Micromechanical characterization of casting-induced inhomogeneity in an Al$_{0.8}$CoCrCuFeNi high-entropy alloy. Scr. Mater. 64, 868−871.

Lu, Z.P., Liu, C.T., 2002. A new glass-forming ability criterion for bulk metallic glasses. Acta Mater. 50, 3501−3512.

Lucas, M.S., Mauger, L., Muoz, J.A., Xiao, Y., Sheets, A.O., Semiatin, S.L., et al., 2011. Magnetic and vibrational properties of high-entropy alloys. J. Appl. Phys. 109, 07E307-1− 07E307-4.

Lucas, M.S., Belyea, D., Bauer, C., Bryant, N., Michel, E., Turgut, Z., et al., 2013. Thermomagnetic analysis of FeCoCrxNi alloys: magnetic entropy of high-entropy alloys. J. Appl. Phys. 113, 17A923-1–17A923-4.

Ma, L., Wang, L., Zhang, T., Inoue, A., 2002. Bulk glass formation of Ti–Zr–Hf–Cu–M (M=Fe, Co, Ni) alloys. Mater. Trans. 43, 277–280.

Ma, S., Chen, Z., Zhang, Y., 2013b. Evolution of microstructures and properties of the Al$_x$CrCuFeNi$_2$ high-entropy alloys. Mater. Sci. Forum 745–746, 706–714.

Ma, S.G., Zhang, Y., 2012. Effect of Nb addition on the microstructure and properties of AlCoCrFeNi high-entropy alloy. Mater. Sci. Eng. A 532, 480–486.

Ma, S.G., Zhang, S.F., Gao, M.C., Liaw, P.K., Zhang, Y., 2013a. A successful synthesis of the CoCrFeNiAl$_{0.3}$ single-crystal, high-entropy alloy by Bridgman solidification. JOM 65, 1751–1758.

Manzoni, A., Daoud, H., Mondal, S., Van Smaalen, S., Völkl, R., Glatzel, U., et al., 2013a. Investigation of phases in Al$_{23}$Co$_{15}$Cr$_{23}$Cu$_8$Fe$_{15}$Ni$_{16}$ and Al$_8$Co$_{17}$Cr$_{17}$Cu$_8$Fe$_{17}$Ni$_{33}$ high entropy alloys and comparison with equilibrium phases predicted by Thermo-Calc. J. Alloys Compd. 552, 430–436.

Manzoni, A., Daoud, H., Völkl, R., Glatzel, U., Wanderka, N., 2013b. Phase separation in equiatomic AlCoCrFeNi high-entropy alloy. Ultramicroscopy 132, 212–215.

Massalski, T.B., 1989. Phase diagrams in materials science. Metall. Trans. A 20A, 1295–1323.

Massalski, T.B., 2001. Binary Alloy Phase Diagrams, vols. 1–3. ASM International, Materials Park, OH, USA.

Materials Genome Initiative for Global Competitiveness. National Science and Technology Council of US, 2011.

Miedema, A.R., 1973. Electronegativity parameter for transition metals. Heat Common Met 32, 117–136.

Miedema, A.R., de Chatel, P.F., de Boer, F.R., 1980. Cohesion in alloys - fundamentals of a semi-empirical model. Physica B 100, 1–28.

Miracle, D.B., 2006. The efficient cluster packing model – An atomic structural model for metallic glasses. Acta Mater. 54, 4317–4336.

Miracle, D.B., Miller, J.D., Senkov, O.N., Woodward, C., Uchic, M.D., Tiley, J., 2014. Exploration and development of high entropy alloys for structural applications. Entropy 16, 494–525.

Mishra, A.K., Samal, S., Biswas, K., 2012. Solidification behaviour of Ti–Cu–Fe–Co–Ni high entropy alloys. Trans. Indian Inst. Met., 725–730.

Mondal, K., Murty, B.S., 2005. On the parameters to assess the glass forming ability of liquids. J. Non-Cryst. Solids 351, 1366–1371.

Morinaga, M., Yukawa, N., Ezaki, H., Adachi, H., 1985. Solid solubilities in nickel-based F.C.C. alloys. Philos. Mag. A 51, 247–252.

Mridha, S., Samal, S., Khan, P.Y., Biswas, K., Govind, 2013. Processing and consolidation of nanocrystalline Cu–Zn–Ti–Fe–Cr high-entropy alloys via mechanical alloying. Metall. Mater. Trans. A 44, 4532–4541.

Muggianu, Y.M., Gambino, M., Bros, J.P., 1975. Enthalpies of formation of liquid alloys bismuth-gallium-tin at 723K - choice of an analytical representation of integral and partial thermodynamic functions of mixing for this ternary system. J. Chim. Phys. Phys.- Chim. Biol. 72, 83–88.

Munitz, A., Kaufman, M.J., Chandler, J.P., Kalaantari, H., Abbaschian, R., 2013. Melt separation phenomena in CoNiCuAlCr high entropy alloy containing silver. Mater. Sci. Eng. A 560, 633–642.

Murty, B.S., Ranganathan, S., 1998. Novel materials synthesis by mechanical alloying. Int. Mater. Rev. 43, 101–141.

Murty, B.S., Ranganathan, S., Rao, M.M., 1992. Solid state amorphization in binary Ti–Ni, Ti–Cu and ternary Ti–Ni–Cu system by mechanical alloying. Mater. Sci. Eng. A 149, 231–240.

Nagase, T., Anada, S., Rack, P.D., Noh, J.H., Yasuda, H., Mori, H., et al., 2012. Electron-irradiation-induced structural change in Zr–Hf–Nb alloy. Intermetallics 26, 122–130.

Nagase, T., Anada, S., Rack, P.D., Noh, J.H., Yasuda, H., Mori, H., et al., 2013. MeV electron-irradiation-induced structural change in the BCC phase of Zr–Hf–Nb alloy with an approximately equiatomic ratio. Intermetallics 38, 70–79.

Ng, C., Guo, S., Luan, J., Wang, Q., Lu, J., Shi, S., et al., 2014. Phase stability and tensile properties of Co-free $Al_{0.5}CrCuFeNi_2$ high-entropy alloys. J. Alloys Compd. 584, 530–537.

Olson, G.B., 1997. Computational design of hierarchically structured materials. Science 277, 1237–1242.

Olson, G.B., 2013. Genomic materials design: the ferrous frontier. Acta Mater. 61, 771–781.

Olson, G.B., 2014. Preface to the viewpoint set on: the materials genome. Scr. Mater. 70, 1–2.

Olson, G.B., Kuehmann, C.J., 2014. Materials genomics: from CALPHAD to flight. Scr. Mater. 70, 25–30.

Otto, F., Dlouhý, A., Somsen, C., Bei, H., Eggeler, G., George, E.P., 2013a. The influences of temperature and microstructure on the tensile properties of a CoCrFeMnNi high-entropy alloy. Acta Mater. 61, 5743–5755.

Otto, F., Yang, Y., Bei, H., George, E.P., 2013b. Relative effects of enthalpy and entropy on the phase stability of equiatomic high-entropy alloys. Acta Mater. 61, 2628–2638.

Palumbo, M., Battezzati, L., 2008. Thermodynamics and kinetics of metallic amorphous phases in the framework of the CALPHAD approach. Calphad 32, 295–314.

Pandeya, S.N., Thakkar, D., 2005. Combinatorial chemistry: a novel method in drug discovery and its application. Indian J. Chem. 44B, 335–348.

Pauzi, S.S.M., Darham, W., Ramli, R., Harun, M.K., Talari, M.K., 2013. Effect of Zr addition on microstructure and properties of $FeCrNiMnCoZr_x$ and $Al_{0.5}FeCrNiMnCoZr_x$ high entropy alloys. Trans. Indian Inst. Met. 66, 305–308.

Pettifor, D.G., 1984. A chemical scale for crystal-structure maps. Solid State Commun. 51, 31–34.

Pettifor, D.G., 1988. Structure maps for pseudobinary and ternary phases. Mater. Sci. Technol. 4, 675–691.

Pettifor, D.G., 1996. Phenomenology and theory in structural prediction. J. Phase Equilib. 17, 384–399.

Pi, J.H., Pan, Y., Zhang, L., Zhang, H., 2011. Microstructure and property of AlTiCrFeNiCu high-entropy alloy. J. Alloys Compd. 509, 5641–5645.

Porter, D.A., Easterling, K.E., 1992. Phase Transformations in Metals and Alloys. CRC Press, London, UK.

Pradeep, K.G., Wanderka, N., Choi, P., Banhart, J., Murty, B.S., Raabe, D., 2013. Atomic-scale compositional characterization of a nanocrystalline AlCrCuFeNiZn high-entropy alloy using atom probe tomography. Acta Mater. 61, 4696–4706.

Praveen, S., Murty, B.S., Kottada, R.S., 2012. Alloying behaviour in multi-component AlCoCrCuFe and NiCoCrCuFe high entropy alloys. Mater. Sci. Eng. A 534, 83–89.

Praveen, S., Anupam, A., Sirasani, T., Murty, B.S., Kottada, R.S., 2013a. Characterization of oxide dispersed AlCoCrFe high entropy alloy synthesized by mechanical alloying and spark plasma sintering. Trans. Indian Inst. Met. 66, 369–373.

Praveen, S., Murty, B.S., Kottada, R.S., 2013b. Phase evolution and densification behaviour of nanocrystalline multicomponent high entropy alloys during spark plasma sintering. JOM 65, 1797–1804.

Qiao, J.W., Ma, S.G., Huang, E.W., Chuang, C.P., Liaw, P.K., Zhang, Y., 2011. Microstructural characteristics and mechanical behaviours of AlCoCrFeNi high-entropy alloys at ambient and cryogenic temperatures. Mater. Sci. Forum 688, 419–425.

Qiu, X.W., 2013. Microstructure and properties of AlCrFeNiCoCu high entropy alloy prepared by powder metallurgy. J. Alloys Compd. 555, 246–249.

Qiu, X.W., Liu, C.G., 2013. Microstructure and properties of $Al_2CrFeCoCuTiNi_x$ high-entropy alloys prepared by laser cladding. J. Alloys Compd. 553, 216–220.

Qiu, X.W., Zhang, Y.P., He, L., Liu, C.G., 2013. Microstructure and corrosion resistance of AlCrFeCuCo high entropy alloy. J. Alloys Compd. 549, 195–199.

Qiu, X.W., Zhang, Y.P., Liu, C.G., 2014. Effect of Ti content on structure and properties of $Al_2CrFeNiCoCuTi_x$ high-entropy alloy coatings. J. Alloys Compd. 585, 282–286.

Raghavan, R., Hari Kumar, K.C., Murty, B.S., 2012. Analysis of phase formation in multi-component alloys. J. Alloys Compd. 544, 152–158.

Ranganathan, S., 2003. Alloyed pleasures: multimetallic cocktails. Curr. Sci. 85, 1404–1406.

Ranganathan, S., Inoue, A., 2006. An application of Pettifor structure maps for the identification of pseudo-binary quasicrystalline intermetallics. Acta Mater. 54, 3647–3656.

Ranganathan, S., Srinivasan, S., 2006. A tale of wootz steel. Resonance 11, 67–77.

Rao, B.R., Srinivas, M., Shah, A.K., Gandhi, A.S., Murty, B.S., 2013. A new thermodynamic parameter to predict glass forming ability in iron based multi-component systems containing zirconium. Intermetallics 35, 73–81.

Razuan, R., Jani, N.A., Harun, M.K., Talari, M.K., 2013. Microstructure and hardness properties investigation of Ti and Nb added $FeNiAlCuCrTi_xN$ by high entropy alloys. Trans. Indian Inst. Met. 66, 309–312.

Reed-Hill, R.E., Abbaschian, R., 1994. Physical Metallurgy Principles. PWS Publishing, Boston, USA.

Ren, B., Liu, Z.X., Li, D.M., Shi, L., Cai, B., Wang, M.X., 2010. Effect of elemental interaction on microstructure of CuCrFeNiMn high entropy alloy system. J. Alloys Compd. 493, 148–153.

Ren, B., Liu, Z.X., Cai, B., Wang, M.X., Shi, L., 2012. Aging behaviour of a $CuCr_2Fe_2NiMn$ high-entropy alloy. Mater. Des. 33, 121–126.

Ren, M.X., Li, B.S., Fu, H.Z., 2013. Formation condition of solid solution type of high entropy alloy. Trans Nonferrous Met. Soc. China 23, 991–995.

Samal, S., Mohanty, S., Mishra, A.K., Biswas, K., Govind, B., 2014. "Mechanical behavior of novel suction cast Ti-Cu-Fe-Co-Ni high entropy alloys". Mater. Sci. Forum 790-791, 503–508.

Senkov, O.N., Miracle, D.B., 2001. Effect of the atomic size distribution on glass forming ability of amorphous metallic alloys. Mater. Res. Bull. 36, 2183–2198.

Senkov, O.N., Woodward, C.F., 2011. Microstructure and properties of a refractory $NbCrMo_{0.5}Ta_{0.5}TiZr$ alloy. Mater. Sci. Eng. A 529, 311–320.

Senkov, O.N., Scott, J.M., Senkova, S.V., Miracle, D.B., Woodward, C.F., 2011a. Microstructure and room temperature properties of a high-entropy TaNbHfZrTi alloy. J. Alloys Compd. 509, 6043–6048.

Senkov, O.N., Scott, J.M., Senkova, S.V., Meisenkothen, F., Miracle, D.B., Woodward, C.F., 2012a. Microstructure and elevated temperature properties of a refractory TaNbHfZrTi alloy. J. Mater. Sci. 47, 4062–4074.

Senkov, O.N., Senkova, S.V., Dimiduk, D.M., Woodward, C., Miracle, D.B., 2012b. Oxidation behaviour of a refractory NbCrMo$_{0.5}$Ta$_{0.5}$TiZr alloy. J. Mater. Sci. 47, 6522–6534.

Senkov, O.N., Senkova, S.V., Miracle, D.B., Woodward, C., 2013a. Mechanical properties of low-density, refractory multi-principal element alloys of the Cr–Nb–Ti–V–Zr system. Mater. Sci. Eng. A 565, 51–62.

Senkov, O.N., Senkova, S.V., Woodward, C., Miracle, D.B., 2013b. Low-density, refractory multi-principal element alloys of the Cr–Nb–Ti–V–Zr system: microstructure and phase analysis. Acta Mater. 61, 1545–1557.

Senkov, O.N., Wilks, G.B., Miracle, D.B., Chuang, C.P., Liaw, P.K., 2010. Refractory high-entropy alloys. Intermetallics 18, 1758–1765.

Senkov, O.N., Wilks, G.B., Scott, J.M., Miracle, D.B., 2011b. Mechanical properties of Nb$_{25}$Mo$_{25}$Ta$_{25}$W$_{25}$ and V$_{20}$Nb$_{20}$Mo$_{20}$Ta$_{20}$W$_{20}$ refractory high entropy alloys. Intermetallics 19, 698–706.

Senkov, O.N., Zhang, F., Miller, J.D., 2013c. Phase composition of a CrMo$_{0.5}$NbTa$_{0.5}$TiZr high entropy alloy: comparison of experimental and simulated data. Entropy 15, 3796–3809.

Shaysultanov, D.G., Stepanov, N.D., Kuznetsov, A.V., Salishchev, G.A., Senkov, O.N., 2013. Phase composition and superplastic behaviour of a wrought AlCoCrCuFeNi high-entropy alloy. JOM 65, 1815–1828.

Shechtman, D., Blech, I., Gratias, D., Cahn, J.W., 1984. Metallic phases with long-range orientational order and no translational symmetry. Phys. Rev. Lett. 53, 1951–1953.

Sheng, H.F., Gong, M., Peng, L.M., 2013. Microstructural characterization and mechanical properties of an Al$_{0.5}$CoCrFeCuNi high-entropy alloy in as-cast and heat-treated/quenched conditions. Mater. Sci. Eng. A 567, 14–20.

Shun, T.T., Du, Y.C., 2009. Microstructure and tensile behaviours of FCC Al$_{0.3}$CoCrFeNi high entropy alloy. J. Alloys Compd. 479, 157–160.

Shun, T.T., Hung, C.H., Lee, C.F., 2010a. Formation of ordered/disordered nano particles in FCC high entropy alloys. J. Alloys Compd. 493, 105–109.

Shun, T.T., Hung, C.H., Lee, C.F., 2010b. The effects of secondary elemental Mo or Ti addition in Al$_{0.3}$CoCrFeNi high-entropy alloy on age hardening at 700°C. J. Alloys Compd. 495, 55–58.

Shun, T.T., Chang, L.Y., Shiu, M.H., 2012a. Microstructure and mechanical properties of multi-principal component CoCrFeNiMo$_x$ alloys. Mater. Charact. 70, 63–67.

Shun, T.T., Chang, L.Y., Shiu, M.H., 2012b. Microstructures and mechanical properties of multi-principal component CoCrFeNiTi$_x$ alloys. Mater. Sci. Eng. A 556, 170–174.

Shun, T.T., Chang, L.Y., Shiu, M.H., 2013. Age-hardening of the CoCrFeNiMo$_{0.85}$ high-entropy alloy. Mater. Charact. 81, 92–96.

Sims, C.T., Hagel, W.C., 1972. The Superalloys. John-Wiley & Sons, Inc., New York, NY.

Singh, A.K., Subramaniam, A., 2014. On the formation of disordered solid solutions in multicomponent alloys. J. Alloys Compd. 587, 113–119.

Singh, S., Wanderka, N., Kiefer, K., Siemensmeyer, K., Banhart, J., 2011a. Effect of decomposition of the Cr–Fe–Co rich phase of AlCoCrCuFeNi high entropy alloy on magnetic properties. Ultramicroscopy 111, 619–622.

Singh, S., Wanderka, N., Murty, B.S., Glatzel, U., Banhart, J., 2011b. Decomposition in multicomponent AlCoCrCuFeNi high-entropy alloy. Acta Mater. 59, 182–190.

Smith, C.S., 1963. Four Outstanding Researchers in Metallurgical History. American Society for Testing and Materials, Baltimore MD.

Smith, C.S., 1981. A Search for Structure. MIT Press, Cambridge, MA.

Smith, W.F., Hashemi, J., 2006. Foundation of Materials Science and Engineering. McGraw-Hill.

Sobol, O.V., Andreev, A.A., Gorban', V.F., Krapivka, N.A., Stolbovoi, V.A., Serdyuk, I.V., et al., 2012. Reproducibility of the single-phase structural state of the multi element high-entropy Ti−V−Zr−Nb−Hf system and related superhard nitrides formed by the vacuum-arc method. Tech. Phys. Lett. 38, 616−619.

Spencer, P.J., 2008. A brief history of CALPHAD. Calphad 32, 1−8.

Sriharitha, R., Murty, B.S., Kottada, R.S., 2013. Phase formation in mechanically alloyed $Al_xCoCrCuFeNi$ ($x = 0.45$, 1, 2.5, 5 mol) high entropy alloys. Intermetallics 32, 119−126.

Sriharitha, R., Murty, B.S., Kottada, R.S., 2014. Alloying, thermal stability and strengthening in spark plasma sintered $Al_xCoCrCuFeNi$ high entropy alloys. J. Alloys Compd. 583, 419−426.

Srinivasan, S., Ranganathan, S., 2014. Indian's Legendary Wootz Steel − An Advanced Material of the Ancient World. Universities Press (India) Pvt. Ltd., Hyderabad, India.

Suryanarayana, C., 2001. Mechanical Alloying and Milling. CRC Press, London, UK.

Swalin, R.A., 1972. Thermodynamics of Solids. John Wiley & Sons, Toronto.

Takeuchi, A., Inoue, A., 2000. Calculations of mixing enthalpy and mismatch entropy for ternary amorphous alloys. Mater. Trans. 41, 1372−1378.

Takeuchi, A., Inoue, A., 2005. Classification of bulk metallic glasses by atomic size difference, heat of mixing and period of constituent elements and its application to characterization of the main alloying element. Mater. Trans. 46, 2817−2829.

Takeuchi, A., Inoue, A., 2006. Analyses of characteristics of atomic pairs in ferrous bulk metallic glasses using classification of bulk metallic glasses and pettifor map. J. Optoelectron. Adv. Mater. 8, 1679−1684.

Takeuchi, A., Inoue, A., 2010. Mixing enthalpy of liquid phase calculated by Miedema's scheme and approximated with sub-regular solution model for assessing forming ability of amorphous and glassy alloys. Intermetallics 18, 1779−1789.

Takeuchi, A., Amiya, K., Wada, T., Yubuta, K., Zhang, W., Makino, A., 2013a. Entropies in alloy design for high-entropy and bulk glassy alloys. Entropy 15, 3810−3821.

Takeuchi, A., Chen, N., Wada, T., Yokoyama, Y., Kato, H., Inoue, A., et al., 2011. $Pd_{20}Pt_{20}Cu_{20}Ni_{20}P_{20}$ high-entropy alloy as a bulk metallic glass in the centimeter. Intermetallics 19, 1546−1554.

Takeuchi, A., Murty, B.S., Hasegawa, M., Ranganathan, S., Inoue, A., 2007. Analysis of bulk metallic glass formation using a tetrahedron composition diagram that consists of constituent classes based on blocks of elements in the periodic table. Mater. Trans. 48, 1304−1312.

Takeuchi, A., Wang, J., Chen, N., Zhang, W., Yokoyama, Y., Yubuta, K., et al., 2013b. $Al_{0.5}TiZrPdCuNi$ high-entropy (H-E) alloy developed through $Ti_{20}Zr_{20}Pd_{20}Cu_{20}Ni_{20}$ H-E glassy alloy comprising inter-transition metals. Mater. Trans. 54, 776−782.

Tang, W.Y., Yeh, J.W., 2009. Effect of aluminum content on plasma-nitrided $Al_xCoCrCuFeNi$ high-entropy alloys. Metall. Mater. Trans. A 40, 1479−1486.

Tang, W.Y., Chuang, M.H., Chen, H.Y., Yeh, J.W., 2009. Microstructure and mechanical performance of brand-new $Al_{0.3}CrFe_{1.5}MnNi_{0.5}$ high-entropy alloys. Adv. Eng. Mater. 11, 788−794.

Tang, W.Y., Chuang, M.H., Chen, H.Y., Yeh, J.W., 2010. Microstructure and mechanical performance of new $Al_{0.5}CrFe_{1.5}MnNi_{0.5}$ high-entropy alloys improved by plasma nitriding. Surf. Coat. Technol. 204, 3118−3124.

Tang, W.Y., Chuang, M.H., Lin, S.J., Yeh, J.W., 2012. Microstructures and mechanical performance of plasma-nitrided $Al_{0.3}CrFe_{1.5}MnNi_{0.5}$ high-entropy alloys. Metall. Mater. Trans. A 43, 2390−2400.

Tang, Z., Gao, M.C., Diao, H., Yang, T., Liu, J., Zuo, T., et al., 2013. Aluminum alloying effects on lattice types, microstructures and mechanical behaviour of high-entropy alloys systems. JOM 65, 1848−1858.

Tariq, N.H., Naeem, M., Hasan, B.A., Akhter, J.I., Siddique, M., 2013. Effect of W and Zr on structural, thermal and magnetic properties of AlCoCrCuFeNi high entropy alloy. J. Alloys Compd. 556, 79−85.

Tian, F., Delczeg, L., Chen, N., Varga, L.K., Shen, J., Vitos, L., 2013a. Structural stability of $NiCoFeCrAl_x$ high-entropy alloy from ab initio theory. Phys. Rev. B: Condens. Matter 88, 085128−085132.

Tian, F., Varga, L.K., Chen, N., Delczeg, L., Vitos, L., 2013b. Ab initio investigation of high-entropy alloys of 3d elements. Phys. Rev. B: Condens. Matter 87, 075144-1−075144-8.

Tong, C.J., Chen, M.R., Chen, S.K., Yeh, J.W., Shun, T.T., Lin, S.J., et al., 2005a. Mechanical performance of the $Al_xCoCrCuFeNi$ high-entropy alloy system with multiprincipal elements. Metall. Mater. Trans. A 36, 1263−1271.

Tong, C.J., Chen, Y.L., Chen, S.K., Yeh, J.W., Shun, T.T., Tsau, C.H., et al., 2005b. Microstructure characterization of $Al_xCoCrCuFeNi$ high-entropy alloy system with multiprincipal elements. Metall. Mater. Trans. A 36, 881−893.

Tsai, C.F., Yeh, K.Y., Wu, P.W., Hsieh, Y.F., Lin, P., 2009a. Effect of platinum present in multi-element nano particles on methanol oxidation. J. Alloys Compd. 478, 868−871.

Tsai, C.W., Chen, Y.L., Tsai, M.H., Yeh, J.W., Shun, T.T., Chen, S.K., 2009b. Deformation and annealing behaviours of high-entropy alloy $Al_{0.5}CoCrCuFeNi$. J. Alloys Compd. 486, 427−435.

Tsai, C.W., Lai, S.W., Cheng, K.H., Tsai, M.H., Davison, A., Tsau, C.H., et al., 2012. Strong amorphization of high-entropy AlBCrSiTi nitride film. Thin Solid Films 520, 2613−2618.

Tsai, C.W., Shun, T.T., Yeh, J.W., 2003a. Study on the Deformation Behavior and Microstructure of CuCoNiCrAlxFe High-Entropy Alloys, Master's thesis. National Tsing Hua University, Taiwan.

Tsai, C.W., Tsai, M.H., Yeh, J.W., Yang, C.C., 2010a. Effect of temperature on mechanical properties of $Al_{0.5}CoCrCuFeNi$ wrought alloy. J. Alloys Compd. 490, 160−165.

Tsai, D.C., Shieu, F.S., Chang, S.Y., Yao, H.C., Deng, M.J., 2010b. Structures and characterizations of TiVCr and TiVCrZrY films deposited by magnetron sputtering under different bias powers. J. Electrochem. Soc. 157, K52−K58.

Tsai, D.C., Huang, Y.L., Lin, S.R., Liang, S.C., Shieu, F.S., 2010c. Effect of nitrogen flow ratios on the structure and mechanical properties of (TiVCrZrY)N coatings prepared by reactive magnetron sputtering. Appl. Surf. Sci. 257, 1361−1367.

Tsai, D.C., Chang, Z.C., Kuo, L.Y., Lin, T.J., Lin, T.N., Shieu, F.S., 2013a. Solid solution coating of (TiVCrZrHf)N with unusual structural evolution. Surf. Coat. Technol. 217, 84−87.

Tsai, K.Y., Tsai, M.H., Yeh, J.W., 2013b. Sluggish diffusion in Co−Cr−Fe−Mn−Ni high-entropy alloys. Acta Mater. 61, 4887−4897.

Tsai, M.H., Lai, C.H., Yeh, J.W., Gan, J.Y., 2008a. Effects of nitrogen flow ratio on the structure and properties of reactively sputtered $(AlMoNbSiTaTiVZr)N_x$ coatings. J. Phys. D: Appl. Phys. 41, 235402-1−235402-7.

Tsai, M.H., Tsai, K.Y., Tsai, C.W., Lee, C., Juan, C.C., Yeh, J.W., 2013c. Criterion for sigma phase formation in Cr- and V- containing high-entropy alloys. Mater. Res. Lett. 1, 207−212.

Tsai, M.H., Wang, C.W., Lai, C.H., Yeh, J.W., Gan, J.Y., 2008c. Thermally stable amorphous $(AlMoNbSiTaTiVZr)_{50}N_{50}$ nitride film as diffusion barrier in copper metallization. Appl. Phys. Lett. 92, 052109-1−052109-3.

Tsai, M.H., Wang, C.W., Tsai, C.W., Shen, W.J., Yeh, J.W., Gan, J.W., et al., 2011. Thermal stability and performance of NbSiTaTiZr high-entropy alloy barrier for copper metallization. J. Electrochem. Soc. 158, H1161−H1165.

Tsai, M.H., Yeh, J.W., Gan, J.Y., 2003b. Study on the Evolution of Microstructure and Electric, Properties of Multi-Element High-Entropy Alloy Films, Master's thesis. National Tsing Hua University, Taiwan.

Tsai, M.H., Yeh, J.W., Gan, J.Y., 2008b. Diffusion barrier properties of AlMoNbSiTaTiVZr high-entropy alloy layer between copper and silicon. Thin Solid Films 516, 5527−5530.

Tsai, M.H., Yuan, H., Cheng, G., Xu, W., Jian, W.W., Chuang, M.H., et al., 2013d. Significant hardening due to the formation of a sigma phase matrix in a high entropy alloy. Intermetallics 33, 81−86.

Tsai, M.H., Yuan, H., Cheng, G., Xu, W., Tsai, K.Y., Tsai, C.W., et al., 2013e. Morphology, structure and composition of precipitates in $Al_{0.3}CoCrCu_{0.5}FeNi$ high-entropy alloy. Intermetallics 32, 329−336.

Tsao, L.C., Chen, C.S., Chu, C.P., 2012. Age hardening reaction of the $Al_{0.3}CrFe_{1.5}MnNi_{0.5}$ high entropy alloy. Mater. Des. 36, 854−858.

Tsau, C.H., 2009. Phase transformation and mechanical behaviour of TiFeCoNi alloy during annealing. Mater. Sci. Eng. A 501, 81−86.

Tsau, C.H., Chang, Y.H., 2013. Microstructures and mechanical properties of $TiCrZrNbN_x$ alloy nitride thin films. Entropy 15, 5012−5021.

Tung, C.C., Shun, T.T., Chen, S.W., Yeh, J.W., 2002. Study on the Deformation Microstructure and High Temperature Properties of Cu-Co-Ni-Cr-Al-Fe, Master's thesis. National Tsing Hua University, Taiwan.

Tung, C.C., Yeh, J.W., Shun, T.T., Chen, S.K., Huang, Y.S., Chen, H.C., 2007. On the elemental effect of AlCoCrCuFeNi high-entropy alloy system. Mater. Lett. 61, 1−5.

Turnbull, D., 1969. Under what conditions can a glass be formed? Contemp. Phys. 10, 473−488.

Varalakshmi, S., Kamaraj, M., Murty, B.S., 2008. Synthesis and characterization of nanocrystalline AlFeTiCrZnCu high entropy solid solution by mechanical alloying. J. Alloys Compd. 460, 253−257.

Varalakshmi, S., Appa Rao, G., Kamaraj, M., Murty, B.S., 2010a. Hot consolidation and mechanical properties of nanocrystallineequiatomic AlFeTiCrZnCu high entropy alloy after mechanical alloying. J. Mater. Sci. 45, 5158−5163.

Varalakshmi, S., Kamaraj, M., Murty, B.S., 2010b. Formation and stability of equiatomic and nonequiatomic nanocrystalline CuNiCoZnAlTi high-entropy alloys by mechanical alloying. Metall. Mater. Trans. A 41, 2703−2709.

Varalakshmi, S., Kamaraj, M., Murty, B.S., 2010c. Processing and properties of nanocrystalline CuNiCoZnAlTi high entropy alloys by mechanical alloying. Mater. Sci. Eng. A 527, 1027−1030.

Venugopal, T., Murty, B.S., 2011. Nanostructured materials by high energy ball milling. Encycl. Nanosci. Nanotechnol. 19, 1−41.

Villars, P., Brandenburg, K., Berndt, M., LeClair, S., Jackson, A., Pao, Y.H., et al., 2001. Binary, ternary and quaternary compound former /nonformer prediction via Mendeleev number. J. Alloys Compd. 317−318, 26−38.

Villars, P., Cenzual, K., Daama, J., Chen, Y., Iwata, S., 2004. Data-driven atomic environment prediction for binaries using the Mendeleev number: Part 1. Composition AB. J. Alloys Compd. 367, 167–175.

Wang, C., Mo, Z., Tang, J., 2012a. The study about microstructure characterization of AlCoCrTiNiCu$_x$ high entropy alloy system with multi-principal element. Adv. Mater. Res. 399–401, 3–7.

Wang, F.J., Zhang, Y., 2008. Effect of Co addition on crystal structure and mechanical properties of Ti$_{0.5}$CrFeNiAlCo high entropy alloy. Mater. Sci. Eng. A 496, 214–216.

Wang, F.J., Zhang, Y., Chen, G.L., 2009a. Atomic packing efficiency and phase transition in a high entropy alloy. J. Alloys Compd. 478, 321–324.

Wang, F.J., Zhang, Y., Chen, G.L., Davies, H.A., 2009b. Cooling rate and size effect on the microstructure and mechanical properties of AlCoCrFeNi high entropy alloy. J. Eng. Mater. Technol., Trans. ASME 131, 0345011–0345013.

Wang, J., Zheng, Z., Xu, J., Wang, Y., 2014a. Microstructure and magnetic properties of mechanically alloyed FeSiBAlNi(Nb) high entropy alloys. J. Magn. Magn. Mater. 355, 58–64.

Wang, S., 2013a. Atomic structure modeling of multi-principal-element alloys by the principle of maximum entropy. Entropy 15, 5536–5548.

Wang, S., Ye, H., 2011. First-principles studies on the component dependences of high-entropy alloys. Adv. Mater. Res. 338, 380–383.

Wang, S.Q., 2013b. Atomic modeling and simulation of AlCoCrCuFeNi multi-principal-element alloy. Mater. Sci. Forum 749, 479–483.

Wang, W.R., Wang, W.L., Wang, S.C., Tsai, Y.C., Lai, C.H., Yeh, J.W., 2012b. Effects of Al addition on the microstructure and mechanical property of Al$_x$CoCrFeNi high-entropy alloys. Intermetallics 26, 44–51.

Wang, W.R., Wang, W.L., Yeh, J.W., 2014b. Phases, microstructure and mechanical properties of Al$_x$CoCrFeNi high-entropy alloys at elevated temperatures. J. Alloys Compd. 589, 143–152.

Wang, X., Xie, H., Jia, L., Lu, Z., 2012c. Effect of Ti, Al and Cu addition on structural evolution and phase constitution of FeCoNi system equimolar alloys. Mater. Sci. Forum 724, 335–338.

Wang, Y.P., Li, B.S., Ren, M.X., Yang, C., Fu, H.Z., 2008. Microstructure and compressive properties of AlCrFeCoNi high entropy alloy. Mater. Sci. Eng. A 491, 154–158.

Weeber, A.W., Bakker, H., 1988. Amorphization by ball milling. A review. Physica B: Phys. Condens. Matter 153, 93–135.

Welk, B.A., Williams, R.E.A., Viswanathan, G.B., Gibson, M.A., Liaw, P.K., Fraser, H.L., 2013. Nature of the interfaces between the constituent phases in the high entropy alloy CoCrCuFeNiAl. Ultramicroscopy 134, 193–199.

Wen, L.H., Kou, H.C., Li, J.S., Chang, H., Xue, X.Y., Zhou, L., 2009. Effect of aging temperature on microstructure and properties of AlCoCrCuFeNi high-entropy alloy. Intermetallics 17, 266–269.

Widom, M., Huhn, W.P., Maiti, S., Steurer, S., 2013. Hybrid Monte Carlo/molecular dynamics simulation of a refractory metal high entropy alloy. Metall. Mater. Trans. 45A, 196–200.

Wu, J.M., Lin, S.J., Yeh, J.W., Chen, S.K., Huang, Y.S., Chen, H.C., 2006. Adhesive wear behaviour of Al$_x$CoCrCuFeNi high-entropy alloys as a function of aluminum content. Wear 261, 513–519.

Xiang, X.D., Sun, X.D., Briceno, G., Lou, Y.L., Wang, K.A., Chang, H.Y., et al., 1995. A combinatorial approach to materials discovery. Science 268, 1738–1740.

Xie, L., Brault, P., Thomann, A.L., Bauchire, J.M., 2013. AlCoCrCuFeNi high entropy alloy cluster growth and annealing on silicon: a classical molecular dynamics simulation study. Appl. Surf. Sci. 285, 810–816.

Yang, H.H., Tsai, W.T., Kuo, J.C., Yang, C.C., 2011. Solid/liquid interaction between a multi-component FeCrNiCoMnAl high entropy alloy and molten aluminum. J. Alloys Compd. 509, 8176–8182.

Yang, T.H., Huang, R.T., Wu, C.A., Chen, F.R., Gan, J.Y., Yeh, J.W., et al., 2009. Effect of annealing on atomic ordering of amorphous ZrTaTiNbSi alloy. Appl. Phys. Lett. 95, 241905-1–241905-3.

Yang, X., Zhang, Y., 2012. Prediction of high-entropy stabilized solid-solution in multi-component alloys. Mater. Chem. Phys. 132, 233–238.

Yao, C.Z., Zhang, P., Liu, M., Li, G.R., Ye, J.Q., Liu, P., et al., 2008. Electrochemical preparation and magnetic study of Bi–Fe–Co–Ni–Mn high entropy alloy. Electrochim. Acta 53, 8359–8365.

Ye, G., Wua, B., Zhang, C., Chen, T., Lin, M., Xie, Y., et al., 2012. Study of solidification microstructures of multi-principal high-entropy alloy FeCoNiCrMn by using experiments and simulation. Adv. Mater. Res. 399–401, 1746–1749.

Ye, X., Ma, M., Liu, W., Li, L., Zhong, M., Liu, Y., et al., 2011. Synthesis and characterization of high-entropy alloy $Al_xFeCoNiCuCr$ by laser cladding. Adv. Mater. Sci. Eng., 485942–485949.

Yeh, J.W., 2006. Recent progress in high-entropy alloys. Ann. Chim. Sci. Mat. 31, 633–648.

Yeh, J.W., 2013a. Alloy design strategies and future trends in high-entropy alloys. J. Met. 65, 1759–1771.

Yeh, J.W., 2013b. Future trends of high-entropy alloys. High-Value Metals Forum, MRS-T Annual Meeting-2013, Jhongli, Taiwan.

Yeh, J.W., Chang, S.Y., Hong, Y.D., Chen, S.K., Lin, S.J., 2007a. Anomalous decrease in X-ray diffraction intensities of Cu–Ni–Al–Co–Cr–Fe–Si alloy systems with multi-principal elements. Mater. Chem. Phys. 103, 41–46.

Yeh, J.W., Chen, S.K., Gan, J.Y., Lin, S.J., Chin, T.S., Shun, T.T., et al., 2004a. Formation of simple crystal structures in Cu–Co–Ni–Cr–Al–Fe–Ti–V alloys with multiprincipal metallic elements. Metall. Mater. Trans. A 35, 2533–2536.

Yeh, J.W., Chen, Y.L., Lin, S.J., Chen, S.K., 2007b. High-entropy alloys - A new era of exploitation. Mater. Sci. Forum 560, 1–9.

Yeh, J.W., Chen, S.K., Lin, S.J., Gan, J.Y., Chin, T.S., Shun, T.T., et al., 2004b. Nanostructured high-entropy alloys with multiple principal elements: novel alloy design concepts and outcomes. Adv. Eng. Mater. 6, 299–303.

Yeh, Y.A., Tsai, M.H., Yeh, J.W., 2012. High-entropy carbides based on high-entropy alloys. 2012 TMS Annual Meeting Bulk Metallic Glasses IX, Orlando.

Yu, Y., Liu, W.M., Zhang, T.B., Li, J.S., Wang, J., Kou, H.C., et al., 2013. Microstructure and tribological properties of $AlCoCrFeNiTi_{0.5}$ high-entropy alloy in hydrogen peroxide solution. Metall. Mater. Trans. 45A, 1–7.

Yue, T.M., Xie, H., Lin, X., Yang, H., Meng, G., 2013. Microstructure of laser re-melted AlCoCrCuFeNi high entropy alloy coatings produced by plasma spraying. Entropy 15, 2833–2845.

Yue, T.M., Xie, H., Lin, X., Yang, H.O., Meng, G.H., 2014. Solidification behaviour in laser cladding of AlCoCrCuFeNi high-entropy alloy on magnesium substrates. J. Alloys Compd. 587, 588–593.

Yuhu, F., Yunpeng, Z., Hongyan, G., Huimin, S., Li, H., 2013. $AlNiCrFe_xMo_{0.2}CoCu$ high entropy alloys prepared by powder metallurgy. Rare Met. Mater. Eng. 42, 1127–1129.

Zaddach, A.J., Niu, C., Koch, C.C., Irving, D.L., 2013. Mechanical properties and stacking fault energies of NiFeCrCoMn high-entropy alloy. J. Mater. Met. Mater. Sci. 65, 1780–1789.

Zhai Q.Y., and Xu J.F., (2011) China Patent CN101554686B.

Zhang, C., Zhang, F., Chen, S., Cao, W., 2012a. Computational thermodynamics aided high-entropy alloy design. JOM 64, 839–845.

Zhang, H., He, Y., Pan, Y., 2013a. Enhanced hardness and fracture toughness of the laser-solidified FeCoNiCrCuTiMoAlSiB$_{0.5}$ high-entropy alloy by martensite strengthening. Scr. Mater. 69, 342–345.

Zhang, H., He, Y., Pan, Y., He, Y., Shin, K., 2010a. Synthesis and characterization of NiCoFeCrAl$_3$ high entropy alloy coating by laser cladding. Adv. Mater. Res. 97–101, 1408–1411.

Zhang, H., He, Y.Z., Pan, Y., Pei, L.Z., 2011a. Phase selection, microstructure and properties of laser rapidly solidified FeCoNiCrAl$_2$Si coating. Intermetallics 19, 1130–1135.

Zhang, H., Pan, Y., He, Y., 2011b. Effects of annealing on the microstructure and properties of Fe$_6$NiCoCrAlTiSi high-entropy alloy coating prepared by laser cladding. J. Therm. Spray Technol. 20, 1049–1055.

Zhang, H., Pan, Y., He, Y., Jiao, H., 2011c. Microstructure and properties of Fe$_6$NiCoSiCrAlTi high-entropy alloy coating prepared by laser cladding. Appl. Surf. Sci. 257, 2259–2263.

Zhang, H., Pan, Y., He, Y.Z., 2011d. Grain refinement and boundary misorientation transition by annealing in the laser rapid solidified Fe$_6$NiCoCrAlTiSi multicomponent ferrous alloy coating. Surf. Coat. Technol. 205, 4068–4072.

Zhang, H., Pan, Y., He, Y.Z., 2011e. Synthesis and characterization of FeCoNiCrCu high-entropy alloy coating by laser cladding. Mater. Des. 32, 1910–1915.

Zhang, K., Fu, Z., 2012. Effects of annealing treatment on properties of CoCrFeNiTiAl$_x$ multi-component alloys. Intermetallics 28, 34–39.

Zhang, K., Fu, Z., Zhang, J., Wang, W., Wang, H., Wang, Y., et al., 2009c. Characterization of nanocrystalline CoCrFeNiCuAl high-entropy alloy powder processed by mechanical alloying. Mater. Sci. Forum 620–622, 383–386.

Zhang, K.B., Fu, Z.Y., Zhang, J.Y., Wang, W.M., Wang, H., Wang, Y.C., et al., 2009a. Microstructure and mechanical properties of CoCrFeNiTiAl$_x$ high-entropy alloys. Mater. Sci. Eng. A 508, 214–219.

Zhang, K.B., Fu, Z.Y., Zhang, J.Y., Shi, J., Wang, W.M., Wang, H., et al., 2009b. Nanocrystalline CoCrFeNiCuAl high-entropy solid solution synthesized by mechanical alloying. J. Alloys Compd. 485, L31–L34.

Zhang, K.B., Fu, Z.Y., Zhang, J.Y., Shi, J., Wang, W.M., Wang, H., et al., 2010b. Annealing on the structure and properties evolution of the CoCrFeNiCuAl high-entropy alloy. J. Alloys Compd. 502, 295–299.

Zhang, K.B., Fu, Z.Y., Zhang, J.Y., Wang, W.M., Lee, S.W., Niihara, K., 2010c. Characterization of nanocrystalline CoCrFeNiTiAl high-entropy solid solution processed by mechanical alloying. J. Alloys Compd. 495, 33–38.

Zhang, L.C., Shen, Z.Q., Xu, J., 2003. Glass formation in a (Ti,Zr,Hf)–(Cu,Ni,Ag)–Al high-order alloy system by mechanical alloying. J. Mater. Res. 18, 2141–2149.

Zhang, L.C., Kim, K.B., Yu, P., Zhang, W.Y., Kunz, U., Eckert, J., 2007. Amorphization in mechanically alloyed (Ti, Zr, Nb)–(Cu, Ni)–Al equiatomic alloys. J. Alloys Compd. 428, 157–163.

Zhang, Y., Chen, G.L., Gan, C.L., 2010d. Phase change and mechanical behaviours of Ti$_x$CoCrFeNiCu$_{1-y}$Al$_y$ high entropy alloys. J. ASTM Int. 7, 1–8.

Zhang, Y., Ma, S.G., Qiao, J.W., 2012b. Morphology transition from dendrites to equiaxed grains for AlCoCrFeNi high-entropy alloys by copper mold casting and bridgman solidification. Metall. Mater. Trans. A 43, 2625–2630.

Zhang, Y., Yang, X., Liaw, P.K., 2012c. Alloy design and properties optimization of high-entropy alloys. JOM 64, 830–838.

Zhang, Y., Zhou, Y., Hui, X., Wang, M., Chen, G., 2008a. Minor alloying behaviour in bulk metallic glasses and high-entropy alloys. Sci. China, Ser. G 51, 427–437.

Zhang, Y., Zhou, Y.J., Lin, J.P., Chen, G.L., Liaw, P.K., 2008b. Solid-solution phase formation rules for multi-component alloys. Adv. Eng. Mater. 10, 534–538.

Zhang, Y., Zuo, T.T., Cheng, Y.Q., Liaw, P.K., 2013b. High-entropy alloys with high saturation magnetization, electrical resistivity and malleability. Sci. Rep. 3, 1–7.

Zhang, Y., Zuo, T.T., Liao, W.B., Liaw, P.K., 2012d. Processing and properties of high-entropy alloys and micro- and nano-wires. Electro Chem. Soc. Trans. 41 (30), 49–60.

Zhang, Y., Zuo, T.T., Tang, Z., Gao, M.C., Dahmen, K.A., Liaw, P.K., et al., 2014. Microstructures and properties of high-entropy alloys. Prog. Mater. Sci. 61, 1–93.

Zhao, F., Cao, W., Ge, C., Tan, Y., Zhang, Y., Fei, Q., 2009. Research on laser engineered net shaping of thick-wall nickel-based alloy parts. Rapid Prototyping J. 15, 24–28.

Zheng, B., Liu, Q., Zhang, L., 2013. Microstructure and properties of MoFeCrTiW high-entropy alloy coating prepared by laser cladding. Adv. Mater. Res. 820, 63–66.

Zhou, Y.J., Zhang, Y., Wang, Y.L., Chen, G.L., 2007a. Microstructure and compressive properties of multicomponent $Al_x(TiVCrMnFeCoNiCu)_{100-x}$ high-entropy alloys. Mater. Sci. Eng. A 454–455, 260–265.

Zhou, Y.J., Zhang, Y., Wang, Y.L., Chen, G.L., 2007b. Solid solution alloys of $AlCoCrFeNiTi_x$ with excellent room-temperature mechanical properties. Appl. Phys. Lett. 90, 181904-1–181904-3.

Zhou, Y.J., Zhang, Y., Kim, T.N., Chen, G.L., 2008a. Microstructure characterizations and strengthening mechanism of multi-principal component $AlCoCrFeNiTi_{0.5}$ solid solution alloy with excellent mechanical properties. Mater. Lett. 62, 2673–2676.

Zhou, Y.J., Zhang, Y., Wang, F.J., Chen, G.L., 2008b. Phase transformation induced by lattice distortion in multiprincipal component $CoCrFeNiCu_xAl_{1-x}$ solid-solution alloys. Appl. Phys. Lett. 92, 241917-1–241917-3.

Zhou, Y.J., Zhang, Y., Wang, F.J., Wang, Y.L., Chen, G.L., 2008c. Effect of Cu addition on the microstructure and mechanical properties of $AlCoCrFeNiTi_{0.5}$ solid-solution alloy. J. Alloys Compd. 466, 201–204.

Zhu, C., Lu, Z.P., Nieh, T.G., 2013. Incipient plasticity and dislocation nucleation of FeCoCrNiMn high-entropy alloy. Acta Mater. 61, 2993–3001.

Zhu, G., Liu, Y., Ye, J., 2014. Early high-temperature oxidation behavior of Ti(C,N)-based cermets with multi-component AlCoCrFeNi high-entropy alloy binder. Int. J. Refract. Met. Hard Mater. 44, 35–41.

Zhu, J.M., Fu, H.M., Zhang, H.F., Wang, A.M., Li, H., Hu, Z.Q., 2010a. Microstructures and compressive properties of multicomponent $AlCoCrFeNiMo_x$ alloys. Mater. Sci. Eng. A 527, 6975–6979.

Zhu, J.M., Fu, H.M., Zhang, H.F., Wang, A.M., Li, H., Hu, Z.Q., 2010b. Synthesis and properties of multiprincipal component $AlCoCrFeNiSi_x$ alloys. Mater. Sci. Eng. A 527, 7210–7214.

Zhuang, Y.X., Liu, W.J., Chen, Z.Y., Xue, H.D., He, J.C., 2012. Effect of elemental interaction on microstructure and mechanical properties of FeCoNiCuAl alloys. Mater. Sci. Eng. A 556, 395–399.

Zhuang, Y.X., Xue, H.D., Chen, Z.Y., Hu, Z.Y., He, J.C., 2013. Effect of annealing treatment on microstructures and mechanical properties of FeCoNiCuAl high entropy alloys. Mater. Sci. Eng. A 572, 30–35.

Zuo, T.T., Ren, S.B., Liaw, P.K., Zhang, Y., 2013. Processing effects on the magnetic and mechanical properties of $FeCoNiAl_{0.2}Si_{0.2}$ high entropy alloy. Int. J. Miner. Metall. Mater. 20, 549–555.

This appendix gives an idea of the structures of various equiatomic and nonequiatomic HEA solid solutions. The HEAs have been grouped according to their structures.

Table A1.1 HEAs with BCC Structure

Composition	Processing Route	Reference
Equiatomic Alloys		
AlFeTi	MA	Varalakshmi et al. (2008)
CrTiV	MaS	Tsai et al. (2010b)
HfNbZr	Sputtering	Guo et al. (2013c) and Nagase et al. (2013)
AlCrFeTi	MA	Varalakshmi et al. (2008)
MoNbTaW	AM	Senkov et al. (2010, 2011b) and Widom et al. (2013)
AlCoCrFeNi	MA	Ji et al. (2014)
AlCoCrFeNi	SC	Wang et al. (2009b) and Qiao et al. (2011)
AlCoCrFeNi	AM	Wang et al. (2008) and Zhou et al. (2008b)
AlCoCrFeNi	BS	Zhang et al. (2012b)
AlCoCrNiSi	GTAW cladding	Lin and Cho (2008)
AlCoCrFeNi	Electro-spark deposition	Li et al. (2013c)
AlCoFeNiTi	AM	Wang et al. (2012c)
AlCrFeTiZn	MA	Varalakshmi et al. (2008)
HfNbTaTiZr	AM and HIP	Senkov et al. (2011a, 2012a)
HfNbTiVZr	AM	Sobol et al. (2012)
MoNbTaVW	AM	Senkov et al. (2010, 2011b)
AlCoCrCuFeNi	MA	Tariq et al. (2013) and Zhang et al. (2009b,c)
AlCoCrCuFeNi	Splat quenching	Singh et al. (2011b)
AlCrCuFeTiZn	MA	Varalakshmi et al. (2008, 2010a)
AlCoCuNiTiZn	MA	Varalakshmi et al. (2010b,c)
CrCoFeMnNiW	MA	Varalakshmi et al. (2008)
AlCoCrCuFeNiW	MA	Tariq et al. (2013)
AlCoCrCuFeNiWZr	MA	Tariq et al. (2013)

(Continued)

Table A1.1 (Continued)

Composition	Processing Route	Reference
Nonequiatomic Alloys		
$AlCoCrCu_{0.5}Ni$	Sputtering	Yeh et al. (2004b)
$AlCoCrCu_{0.5}Ni$	AM	Yeh et al. (2004b)
$Al_xCoCrFeNi$ ($x = 0.9-2$)	AM	Wang et al. (2012b)
$Al_{0.7}Co_{0.3}CrFeNi$	MA	Chen et al. (2013b)
$Al_{0.5}CoFeNiSi_{0.5}$	AM and SC	Zhang et al. (2012d)
$Al_xCoFeNiSi$ ($x > 0.3$)	AM	Zhang et al. (2013b)
$Al_xCrFe_{1.5}MnNi_{0.5}$ ($x = 0.8-1.2$)	AM	Tsai et al. (2013c)
$Al_{0.5}CrMoNbTi$	AM	Liu et al. (2014)
$MoNbV_{0.5}TiZr$	SC	Zhang et al. (2012c)
$AlCoCrCu_xFeNi$	AM	Zhang et al. (2010d)
$AlCoCrCu_{0.5}FeNi$	AM	Tung et al. (2007)
$AlCoCrCu_{0.25}FeNi$	IC	Zhang et al. (2008b)
$AlCoCrFeMo_xNi$ ($x = 0$ and 0.1)	AM	Zhu et al. (2010a)
$AlCoCrFeNb_xNi$ ($x = 0-0.1$)	AM	Zhang et al. (2012c)
$AlCoCrFeNiSi_x$ ($x = 0-0.8$)	AM	Zhu et al. (2010b)
$Al_{0.5}CrFe_{1.5}MnNi_{0.5}$	AM	Tang et al. (2010)
$Al_{0.5}CrMoNbTiV$	AM	Liu et al. (2014)
$AlCoCrFe_6NiSiTi$	LC	Zhang et al. (2011b,c,d)
$AlCrFeMo_{0.5}NiSiTi$	Plasma spray	Huang et al. (2004)
$AlCoCrFeMo_{0.5}NiSiTi$	Plasma spray	Huang et al. (2004)
$Al_2CoCrCuFeMnNiTiV$	AM	Zhang et al. (2008a)

Table A1.2 HEAs with FCC Structure

Composition	Processing Route	Reference
Equiatomic Alloys		
CoCuNi	MA	Varalakshmi et al. (2010b)
CoFeNi	DC	Guo et al. (2014)
CoFeNi	IM	Singh and Subramaniam (2014)
CoFeNi	AM	Wang et al. (2012c)
CoFeNi	MA	Praveen et al. (2012, 2013b)
CoCrFeNi	DC	Guo et al. (2014)
CoCrCuNi	MA	Durga et al. (2012)
CoCuFeNi	MA	Praveen et al. (2012, 2013b)
CoCuNiZn	MA	Varalakshmi et al. (2010b)
CoFeMnNi	MA	Praveen et al. (2013b)
AlCoCuNiZn	MA	Varalakshmi et al. (2010b)
CoCrCuFeNi	AM	Cui et al. (2011b) and Li et al. (2009)
CoCrCuFeNi	LC	Zhang et al. (2011e) and Cheng et al. (2013)
CoCrFeMnNi	IM	Ye et al. (2012)
CoCrFeMnNi	AM	Bhattacharjee et al. (2014), Gali and George (2013), and Zaddach et al. (2013)
CoCrFeMnNi	DC	Liu et al. (2013), Otto et al. (2013a), and Zhu et al. (2013)
CoCrFeMnNi	MeS	Cantor et al. (2004)
CoCrFeMnNi	MA	Zaddach et al. (2013)
CoCrFeNiTi	AM	Zhang et al. (2009a)
CoCuFeNiTi	AM	Wang et al. (2012c)
CoCuFeNiV	IC	Zhang et al. (2008b)
CrCuFeMoNi	AM	Li et al. (2009)
CrTiVYZr	MaS	Tsai et al. (2010b)
AlCoCrCuFeNi	MaS	Dolique et al. (2009, 2010)
CoCrFeMnNiCu	MeS	Cantor et al. (2004)
CoCrFeMnNiNb	MeS	
CoCrFeMnNiV	MeS	
Nonequiatomic Alloys		
$Al_{0.5}CoCrFeNi$	AM	Lin and Tsai (2011)
$Al_{0.3}CoCrFeNi$	AM and BS	Ma et al. (2013a)
$Al_{0.3}CoCrFeNi$	AM	Shun et al. (2010a,b)
$Al_{0.5}CoCrFeNi$	IC	Zhang et al. (2008b)
$Al_xCoCrFeNi$ $(x = 0-0.3)$	AM	Wang et al. (2014b)
$Al_xCoCrFeNi$ $(x = 0-0.4)$	AM	Wang et al. (2012b)
$Al_xCoCrFeNi$ $(x = 0-0.45)$	AM	Kao et al. (2011)
$Al_xCoCrFeNi$ $(x = 0-0.65)$	AM	Tian et al. (2013a)

(Continued)

Table A1.2 (Continued)

Composition	Processing Route	Reference
$Al_{0.25}CoCrFeNi$	AM	Kao et al. (2009)
$Al_xCoCrFeNi$ ($x = 0-0.375$)	AM	Chou et al. (2009)
$Al_xCoFeNiSi_x$ ($x < 0.3$)	AM	Zhang et al. (2013b)
$Al_xCoFeNiSi_{0.2}$	AM and SC	Zhang et al. (2012d)
$Al_{0.2}CoFeNiSi_{0.2}$	AM and IM	Zuo et al. (2013)
$Al_xCrCuFeNi_2$ ($x \leq 0.5$)	DC	Guo et al. (2013b)
$Al_{0.5}CrCuFeNi_2$	SC	Ma et al. (2013b)
$CoCrCu_{0.5}FeNi$	AM	Hsu et al. (2005)
$CoCrFeMo_{0.3}Ni$	AM	Shun et al. (2010a, 2012a)
$CoCrFeNiTi_{0.3}$	AM	Shun et al. (2012b)
$CoCuFeNiSn_x$ ($x = 0-0.05$)	AM	Liu et al. (2012)
$CrCu_2Fe_2Ni_2Mn$	AM	Ren et al. (2010, 2012)
$Cr_2CuFe_2Mn_2Ni_2$	AM	
$CrCuFeMn_2Ni_2$	AM	
$AgCoCuFeNi\,Pt_x$	RF Sputtering	Tsai et al. (2009a)
$Al_xCoCrCu_{1-x}FeNi$	AM	Zhang et al. (2010d)
$Al_{0.5}CoCrCuFeNi$	AM	Yeh et al. (2004b)
$Al_{0.5}CoCrCuFeNi$	Sputtering	Chen et al. (2004)
$Al_xCoCrCuFeNi$	RF Sputtering	Yeh et al. (2004a)
$Al_xCoCrCuFeNi$	AM	Tong et al. (2005b)
$Al_xCoCrCuFeNi$	AM	Wu et al. (2006)
$Al_{0.5}CoCrCuFeNi$	AM	Tung et al. (2007)
$Al_xCoCrCu_{1-x}FeNi$	AM	Zhou et al. (2008b)
$Al_xCoCrCuFeNi$	AM	Tang and Yeh (2009)
$Al_{0.5}CoCrCuFeNi$	AM	Tsai et al. (2009b)
$Al_{0.5}CoCrCu_{0.5}FeNi$	AM	Li et al. (2013a)
$Al_{0.3}CoCrCu_{0.5}FeNi$	AM	Tsai et al. (2013e)
$Al_{0.5}CoCrCu_{0.5}FeNi_2$	IM	Daoud et al. (2013)
$Al_x(CoCrFeMnNi)_{100-x}$ ($x < 8$ at.%)	AM	He et al. (2014)
$Al_{0.3}CoCrFeMo_{0.1}Ni$	AM	Shun et al. (2010a,b)
$Al_xCoCrFeMo_{0.5}Ni$ ($x = 0-0.5$)	AM	Hsu et al. (2013a,b)
$Al_{0.3}CoCrFeNiTi_{0.1}$	AM	Shun et al. (2010b)
$Al_xCo_{1.5}CrFeNiTi_y$	AM	Chuang et al. (2011)
$Al_xCoCrFeNiTi_{0.5}$ ($x = 0, 0.2$)	AM	Dong et al. (2013b)
$AlCoCu_xNiTiZn$	MA	Varalakshmi et al. (2010b)
$Al_{0.5}CoCrFeMnNi$	AM	Pauzi et al. (2013)
$Al_xCoCrCu_{1-x}FeNiTi_{0.5}$ ($x = 0, 0.25$)	SC	Wang et al. (2009a)
$Al_{0.5}CoCrCuFeNiV_x$ ($x < 0.4$)	AM	Chen et al. (2006a)
$Co_{1.5}CrFeNi_{1.5}Ti_{0.5}Mo_x$ ($x = 0,0.1$)	AM	Chou et al. (2010b)

Table A1.3 HEAs Showing More Than One Solid Solution

Composition	Processing Route	Major Phase	Minor Phase	Reference
Equiatomic Compositions				
AlCuNi	AM	BCC	FCC	Yeh et al. (2007a)
CrFeNi	IM	FCC	BCC	Singh and Subramaniam (2014)
AlCoCuNi	AM	BCC	FCC	Yeh et al. (2007a)
CoCrFeNi	MA	FCC	BCC	Praveen et al. (2012)
CoFeNiTi	AM	HCP	FCC	Tsau (2009)
CoCrFeNi	MA	FCC	BCC	Praveen et al. (2013b)
CoCuFeNi	IM	FCC1	FCC2	Singh and Subramaniam, (2014)
AlCoCrCuNi	GTAW cladding	FCC	BCC	Lin and Cho (2009)
AlCoCrCuNi	AM	BCC	FCC	Yeh et al. (2007a)
AlCoCrCuFe	LC	FCC	BCC	Qiu et al. (2013)
AlCoCuFeNi	AM	BCC	FCC	Zhuang et al. (2013)
AlCoCuFeNi	SC	BCC	FCC	Zhuang et al. (2012)
AlCrCuFeNi	AM	BCC	FCC	Li et al. (2009)
CoCrCuFeNi	MA	FCC	BCC	Praveen et al. (2012, 2013b)
CrCuFeMnNi	AM	FCC	BCC	Li et al. (2009) and Ren et al. (2010, 2012)
AlCoCrCuFeNi	Plasma spraying	BCC	FCC	Yue et al. (2013)
AlCoCrCuFeNi	Powder metallurgy	FCC	BCC	Qiu (2013)
AlCoCrCuFeNi	IM	BCC	FCC1 + FCC2	Kuznetsov et al. (2012, 2013) and Singh et al. (2011b)
AlCoCrCuFeNi	SC	BCC	FCC	Zhuang et al. (2012)
AlCoCrCuFeNi	LC	BCC	FCC	Yue et al. (2014)
AlCoCrCuFeNi	AM	BCC	FCC	Tung et al. (2007), Yeh et al. (2007a), and Wen et al. (2009)

(Continued)

Table A1.3 (Continued)

Composition	Processing Route	Major Phase	Minor Phase	Reference
AlCoCrCuFeNi	AM	FCC	BCC	Wu et al. (2006)
AlCoCrCuFeNi	MaS	BCC	FCC	Dolique et al. (2010)
AlCoCuFeNiTi	AM	BCC	FCC	Wang et al. (2012c)
AlCoCuFeNiTi	SC	BCC	FCC	Zhuang et al. (2012)
AlCrCuFeNiZn	MA	BCC	FCC	Pradeep et al. (2013) and Koundinya et al. (2013)
AlCoCrCuFeMnNi	AM	BCC	FCC	Li et al. (2008a)
AlCoCrCuFeNiTi	AM	BCC1	BCC2, FCC	
AlCoCrCuFeNiV	AM	BCC	FCC	
AlCoCrCuFeNiSi	AM	BCC	FCC	Yeh et al. (2007a)
AlCrCuFeNbNiTi	AM	FCC1 + FCC2	BCC	Razuan et al. (2013)
CoCrFeNiCuAlTiXVMo (X = Zn, Mn)	MA	BCC	FCC	Fazakas et al. (2013)
Nonequiatomic Compositions				
$NbTiV_2Zr$	AM	BCC2	BCC1 + BCC3	Senkov et al. (2013b)
$Al_xCoCrFeNi$	AM	$x = 0.5-0.75$: FCC	BCC	Chou et al. (2009)
		$x = 0.875-1$: BCC	FCC	
$Al_xCoCrFeNi$ ($x = 0.5, 0.75$)	AM	FCC	BCC	Kao et al. (2009)
$Al_{0.3}CrFe_{1.5}MnNi_{0.5}$	AM	FCC	BCC	Tang et al. (2009)
$Al_{0.3}CrFe_{1.5}MnNi_{0.5}$	AM	FCC	BCC	Chen et al. (2010a)
$Al_{0.5}CrCuFeNi_2$	DC	FCC1	FCC2	Ng et al. (2014)
$Al_{0.3}CrFe_{1.5}MnNi_{0.5}$	AM	BCC	FCC	Tang et al. (2012)
$Al_{0.3}CrFe_{1.5}MnNi_{0.5}$	AM	FCC1	BCC + FCC2	Tsao et al. (2012)
$Al_xCoCrFeNi$ ($x = 0.5-0.8$)	AM	FCC	BCC	Wang et al. (2012b)
$Al_{0.75}CoCrCu_{0.25}FeNiTi_{0.5}$	SC	BCC	BCC	Wang et al. (2009a)
$Al_{0.3}CoFeNiSi_{0.3}$	AM	FCC	BCC	Zhang et al. (2013b)
$Al_xCoCrFeNi$ ($x = 0.45-0.85$)	AM	FCC	BCC	Kao et al. (2011)
$Al_{0.3}CrFe_{1.5}MnNi_{0.5}$	IM	FCC	BCC	Chuang et al. (2013)
$AlCrCuFeNi_2$	SC	FCC	BCC	Ma et al. (2013b)

(Continued)

Table A1.3 (Continued)

Composition	Processing Route	Major Phase	Minor Phase	Reference
$Al_{0.3}CoFeNiSi_{0.3}$	AM	FCC	BCC	Zhang et al. (2013b)
$Al_{0.5}CoCrCuFeNiV_x$ ($x = 0.4-2$)	AM	FCC	BCC	Chen et al. (2006a)
$Co_{0.5}CrFeNiTi_{0.5}$	MA	BCC	FCC	Fu et al. (2013b)
$CrCu_xFeTiZn_y$ ($x/y = 1/0, 3/1,$ 1 and $x + y = 40$)	MA	FCC	BCC	Mridha et al. (2013)
$Cr_2Cu_2FeNi_2Mn$	AM	FCC	BCC	Ren et al. (2010, 2012)
$Cr_2Cu_2FeNiMn_2$	AM	FCC	BCC	
$CrCu_2Fe_2NiMn_2$	AM	FCC	BCC	
Cr_2CuFe_2NiMn	AM	FCC	BCC	
$Al_{0.8}CoCrCuFeNi$	AM	FCC	BCC	Tong et al. (2005b)
$AlCo_{0.5}CrCuFeNi$	AM	FCC	BCC	Tung et al. (2007)
$AlCoCr_{0.5}CuFeNi$	AM	FCC	BCC	
$AlCoCrCuFe_{0.5}Ni$	AM	FCC	BCC	
$AlCoCrCuFeNi_{0.5}$	AM	FCC	BCC	
$Al_{0.75}CoCrCu_{0.25}FeNi$	AM	BCC	FCC	Zhou et al. (2008b)
$AlCoCrFeNiTi_x$ ($x = 0.5, 1$)	IC	BCC1	BCC2	Zhou et al. (2007b)
$AlCo_xCrFeNiTi_{0.5}$	SC	$x = 1$: BCC1	BCC2	Wang and Zhang (2008)
		$x = 1.5, 2, 3$: BCC	FCC	
$AlCoCu_xNiTiZn$	MA	$x = 0$: BCC	FCC	Varalakshmi et al. (2010b)
		$x = 8.33$: BCC	FCC	
		$x = 50$: FCC	BCC	
$Al_xCoCrCu_{1-x}FeNi$ ($x = 0.25-0.75$)	AM	FCC	BCC	Zhang et al. (2010d)
$Al_{0.5}C_{0.2}Co_{0.3}CrFeNi$	MA	BCC	FCC	Fang et al. (2014)
$Al_x(CoCrFeMnNi)_{100-x}$ ($x = 8-16$ at.%)	AM	FCC	BCC	He et al. (2014)
$AlCoCrCu_xNiTi$ ($x = 0.5-0.8$)	AM	FCC	BCC	Wang et al. (2012a)
$Al_{0.8}CoCrCuFeNi$	DC	FCC1 + FCC2	BCC	Liu et al. (2011)
$Al_xCoCrCuFeNi$ ($x = 1-2$)	LC	BCC	FCC	Ye et al. (2011)
$Al_xCoCrFeNiTi_{0.5}$ ($x = 0.8-1$)	AM	BCC1	BCC2	Dong et al. (2013b)
$CoNiFeCrAl_{0.6}Ti_{0.4}$	MA	BCC	FCC	Fu et al. (2013a)
$Al_xFeCoNiCrCu_{0.5}$	AM	$x = 1$: BCC	FCC1	Li et al. (2013a)
		$x = 1.5$: BCC	FCC2	
$Al_{0.5}CoCrFeCuNi$	IM	FCC1	FCC2 (Cu rich)	Sheng et al. (2013)

(Continued)

Table A1.3 (Continued)

Composition	Processing Route	Major Phase	Minor Phase	Reference
$Al_xCoCrCuFeNi$ ($x = 0.45, 1$)	MA	FCC1 + FCC2	BCC	Sriharitha et al. (2013)
$CoCrFeMnNiZr_x$ ($x = 0-0.3$)	AM	BCC	FCC	Pauzi et al. (2013)
$AlCuCrFeNiTi_x$	AM	$x = 0$: FCC1	BCC1	Razuan et al. (2013)
		$x = 0.5, 1$: BCC1 + FCC1	BCC2	
		$x = 1.5$: BCC1 + BCC2	FCC1	
$AlCuCrFeNiNb_x$	AM	$x = 0$: FCC1	BCC1	
		$x = 0.5$: FCC1 + FCC2	BCC1	
		$x = 1, 1.5$: FCC1 + FCC2 + BCC1	BCC2	
$Al_2CoCrFeMo_{0.5}Ni$	AM	BCC2	BCC1	Hsu et al. (2013a,b)
$AlCoCrCuFe_2Mo_{0.2}Ni$	MA	BCC	FCC	Yuhu et al. (2013)
$AlCoCrCu_xFeNiTi_{0.5}$ ($x = 0$, 0.25, 0.5)	IC	BCC1	BCC2	Zhou et al. (2008c)
$Al_{0.5}CoCrCu_{0.5}FeNiTi_{0.5}$	SC	FCC	BCC	Wang et al. (2009a)
$Al_2CoCrCuFeNi$	LC	BCC	FCC	Qiu et al. (2014)
$AlCoCrCuFeNiTi_x$	AM	$x = 0.5$: BCC1 + FCC	BCC2	Hongfei et al. (2011)
		$x = 1$: BCC1 + FCC	BCC2	
$Al_2CoCrFeNi_xTi$ ($x = 0-1.0$)	LC	FCC	BCC	Qiu and Liu (2013)
$Al_{0.5}CoCrFeMnNiZr_x$ ($x = 0.1-0.3$)	AM	BCC	FCC	Pauzi et al. (2013)
$Al_{11.1}(CoCrCuFeMnNiTiV)_{88.9}$	IM	FCC	BCC	Zhou et al. (2007a)
$CoCu_yFeNiTi_x$	AM	FCC1 (Cu rich) + FCC2 (Co rich) ($x = 1/3, 3/7, 3/5$)	BCC (β-Ti rich) ($x = 3/5$)	Mishra et al. (2012)
$CoCu_yFeNiTi_x$ ($x/y = 1/3, 3/7, 3/5, 9/11, 11/9, 3/2$)	SC	FCC1	FCC2	Samal et al. (2014)
$AlCoCrFeNiTi_x$ ($x = 0-2$)	AM	BCC1	BCC2	Zhang et al. (2008a)
$Al_xCoCrFeNiTi$ ($x = 1-2$)	AM	BCC	B2	Zhang et al. (2009a)

AM, arc melting; IM, induction melting; IC, injection casting; SC, suction casting; BS, Bridgman solidification; MeS, melt spinning; LENS, laser engineered net shaping; LC, laser cladding; MA, mechanical alloying; MaS, magnetron sputtering.

This appendix gives an idea of the intermetallic compounds/intermediate phases and metallic glasses obtained in various equiatomic and nonequiatomic HEAs. The HEAs have been grouped according the nature of the major intermetallic/intermediate phase observed in them.

Table A2.1 B2 Phase in HEAs

Composition	Processing Route	Major Phase	Minor Phase	Reference
AlCrFeNi	IM	BCC	B2	Singh and Subramaniam (2014)
AlCoCrCuNi	AM	B2	FCC	Hsu et al. (2007)
AlCoCrCuNi	AM	BCC	B2	Munitz et al. (2013)
AlCoCrFeNi	AM	BCC	B2	Hsu et al. (2013a)
AlCoCrFeNi	AM	B2	–	Wang et al. (2008)
AlCoCrFeNi	IM	B2	σ	Manzoni et al. (2013b)
AlCoCrFeNi	SC	B2	–	Qiao et al. (2011)
AlCoCrFeNi	BS	B2	–	Zhang et al. (2012b)
AlCoCrFeNi	Electro-spark deposition	B2	–	Li et al. (2013c)
AlCoCuFeNi	IM	FCC1	FCC2 + B2	Singh and Subramaniam (2014)
AlCoCrNiW	GTAW cladding	B2	W	Lin and Cho (2008)
AlCoCrCuFeNi	AM	B2	FCC	Hsu et al. (2007) and Zhang et al. (2010b)
AlCoCrCuFeNi	IM	B2 + BCC + FCC	Cu	Shaysultanov et al. (2013)
AlCoCrCuFeNi	LENS	B2	BCC	Welk et al. (2013)
AlCoCrFeNiTi	AM	B2	BCC + FCC	Zhang et al. (2009a)
AlCoCrFeNiTi	MA	B2	FCC	Zhang et al. (2010c)
$Al_{0.3}CoCrFeNi$	AM	FCC	B2	Shun and Du (2009)
$Al_xCoCrFeNi$ $(x = 1-3)$	AM	B2	–	Li et al. (2010a)
$Al_xCoCrFeNi$	AM	$x = 0.5-0.7$: FCC	B2	Wang et al. (2014b)
		$x = 0.9-1.8$: B2	–	

(Continued)

Table A2.1 (Continued)

Composition	Processing Route	Major Phase	Minor Phase	Reference
$Al_xCoCrFeNi$	AM	$x = 0.875-1.25$: FCC	B2	Kao et al. (2009)
		$x = 1.5-2$: FCC	B2	
$Al_xCoCrFeNi$ ($x = 1.5-3$)	AM	B2	–	Li et al. (2009)
$Al_2CoCrFeNi$	AM	B2	–	Lucas et al. (2011) and Chen and Kao (2012)
$Al_3CoCrFeNi$	Laser cladding	B2	FCC	Zhang et al. (2010a)
$Al_2CrCuFeNi_2$	AM and SC	BCC	B2	Ma et al. (2013b)
$Al_xCrCuFeNi_2$ ($x = 0.2-2.5$)	IC	$x \geq 0.8$: BCC + B2	FCC	Guo et al. (2013b)
$AlCrFeMo_xNi$	AM	$x = 0-0.5$: BCC	B2	Dong et al. (2013a)
		$x = 0.8-1.0$: B2 + FCC	σ	
$Al_{0.3}CrFe_{1.5}MnNi_{0.5}$	TIG overlay	BCC	B2	Hsieh et al. (2009)
$Al_xCoCrCuFeNi$	AM	$x > 1$: B2	BCC + FCC	Tong et al. (2005b)
$Al_xCoCrCuFeNi$	Sputtering	$x = 0.5-2.5$: B2	FCC + BCC	Yeh et al. (2004a)
		$x > 2.8$: B2	BCC	
$Al_x(CoCrFeMnNi)_{100-x}$	AM	$x > 16$: B2	–	He et al. (2014)
$Al_xCoCrCuFeNi$	MA	$x = 2.5,5$: B2	FCC	Sriharitha et al. (2013)
$Al_2CoCrCuFeNi$	AM	B2	–	Wu et al. (2006)
$Al_2CoCrFeNiSi$	Laser RSP	B2	BCC	Zhang et al. (2011a)
$AlCoCrFeNiTi_{0.5}$	IM	B2	BCC	Yu et al. (2013)
$Al_{0.5}CoCrCuFeNiTi_x$	AM	$x = 0-0.6$: FCC + B2	BCC	Chen et al. (2006b)
$AlCrFeMo_{0.5}NiSiTi$	AM	B2	FCC1 + FCC2	Huang et al. (2004)
$AlCoCrFeMo_{0.5}NiSiTi$	AM	B2 + FCC1	FCC2	Huang et al. (2004)

Table A2.2 L1$_2$ Phase in HEAs

Composition	Processing Route	Major Phase	Minor Phase	Reference
$AlCoCrFeNi$	AM	B2	$L1_2$	Li et al. (2008b)
$Al_{0.5}CoCrCuFeNi$	AM	FCC	$L1_2$	Hemphill et al. (2012)
$Al_{0.5}CoCrCu_{0.5}FeNi_2$	IM	FCC	$L1_2$	Manzoni et al. (2013a)

Table A2.3 Sigma (σ) Phase in HEAs

Composition	Processing Route	Major Phase	Minor Phase	Reference
$CrFe_{1.5}MnNi_{0.5}$	AM	FCC	σ	Lee et al. (2008a) and Tsai et al. (2013c)
AlCoCrFeNi	IM	B2	σ	Manzoni et al. (2013b)
$Al_{0.5}CoCrNiTi_{0.5}$	AM	BCC + B2 + FCC	σ	Lee and Shun (2013)
$AlCrFeMo_xNi$	AM	$x = 0.8-1.0$: B2 + FCC	σ	Dong et al. (2013a)
$Al_{0.3}CrFe_{1.5}MnNi_{0.5}$	AM	BCC + FCC	σ	Tsai et al. (2013c,d)
$Al_xCrFe_{1.5}MnNi_{0.5}$	AM	$x = 0.3$: FCC + BCC	σ	Tsai et al. (2013c)
		$x = 0.5$: BCC	σ	
CoCrCuFeMn	IC	FCC1 + FCC2	σ	Otto et al. (2013b)
CrFeMnNiTi	IC	FCC	σ, Laves	Otto et al. (2013b)
CoFeMnNiV	IC	FCC	σ	Otto et al. (2013b)
CrFeMoTiW	Laser cladding	BCC	σ	Zheng et al. (2013)
$CoCrFeNiMo_x$	AM	$x = 0.5$: FCC	σ	Shun et al. (2012a)
		$x = 0.85$: FCC	σ	
$CoCrFeMo_{0.85}Ni$	AM	FCC	σ	Shun et al. (2013)
$Al_xCoCrFeMo_{0.5}Ni$	AM	$x = 0-0.5$: FCC	σ	Hsu et al. (2013a,b)
		$x = 1-1.5$: BCC	σ	
$AlCo_xCrFeMo_{0.5}Ni$	AM	$x = 0.5-1.5$: BCC	σ	Hsu et al. (2010b)
		$x = 2$: BCC + FCC	σ	
$AlCo_xCrFeMo_{0.5}Ni$	AM	$x = 0-1.5$: BCC + B2	σ	Hsu et al. (2013a)
		$x = 2$: BCC + B2 + FCC	σ	
$AlCoCr_xFeMo_{0.5}Ni$	AM	$x = 0-2$: BCC + B2	σ	Hsu et al. (2011, 2013a)
$AlCoCrFe_xMo_{0.5}Ni$	AM	$x = 0.6-2$: BCC + B2	σ	Hsu et al. (2010a, 2013a)
$AlCoCrFeMo_{0.5}Ni_x$	AM	$x = 0, 0.5, 1$: B2	σ	Juan et al. (2013)
$AlCoCrFeMo_xNi$	AM	$x = 0.5-0.9$: BCC + B2	σ	Hsu et al. (2013a)
$AlCoCrFeMoNi_x$	AM	$x = 0-1$: BCC + B2	σ	Hsu et al. (2013a)
		$x = 1.5-2$: BCC + B2 + FCC	σ	
$Al_{0.5}CoCrCuFeNiV_x$ ($x = 0.6-1$)	AM	FCC + BCC	σ	Chen et al. (2006a)
$Co_{1.5}CrFeNi_{1.5}Ti_{0.5}Mo_x$	AM	FCC	$x > 0.1$: σ	Chou et al. (2010b)
CoCrCuFeMnNiTiV	AM	BCC + FCC	σ	Zhang et al. (2008a)

Table A2.4 Laves Phase in HEAs

Composition	Processing Route	Major Phase	Minor Phase	Reference
CoCrFeNiTi$_x$	AM	$x = 0.5$: FCC	Laves + σ	Shun et al. (2012b)
CoCu$_y$FeNiTi$_x$	AM	FCC1 (Cu rich) + FCC2 (Co rich) + BCC (β-Ti rich)	$x > 3/5$: Laves	Mishra et al. (2012)
CoCu$_y$FeNiTi$_x$	SC	FCC1 + FCC2	$x/y = 1$: Laves	Samal et al. (2014)
CrFeMnNiTi	IC	FCC	σ, Laves	Otto et al. (2013b)
AlCoCrFeNiTi$_x$	AM	BCC1 + BCC2	$x = 3$: Laves	Zhang et al. (2008a)
Al$_x$CoCrFeNiTi	AM	$x = 0.5$: B2 + BCC + FCC	Laves	Zhang et al. (2009a)
AlCoCrFeNb$_x$Ni ($x = 0.25-0.75$)	AM	BCC	Laves	Zhang et al. (2012c)
AlCoCrFeNb$_x$Ni ($x = 0.25-0.75$)	SC	BCC	Laves	Ma and Zhang (2012)
AlCoCrFeNiTi$_{0.5}$	SC	BCC	Laves	Qiao et al. (2011)
Al$_x$CoCrFeNiTi$_{0.5}$	AM	$x = 0.5-0.8$: FCC + BCC	Laves	Dong et al. (2013b)
AlCrCuFeNiTi	AM	BCC1 + BCC2	Laves	Pi et al. (2011)
CoFeMnTi$_x$V$_y$Zr$_z$ ($x = 0.5-2.5$, $y = 0.4-3$, $z = 0.4-3$)	AM	C14 Laves	–	Kao et al. (2010)
CrFeNiTiVZr	LENS	C14 Laves	α-Ti	Kunce et al. (2013)
CrMo$_{0.5}$NbTa$_{0.5}$TiZr	AM	BCC1 + BCC2	Laves	Senkov et al. (2012b, 2013c)
CrMo$_{0.5}$NbTa$_{0.5}$TiZr	AM	BCC1 + BCC2 + FCC	Laves	Senkov and Woodward (2011)

Table A2.5 Other Intermetallic Compounds in HEAs

Composition	Processing Route	Major Phase	Minor Phase	Reference
CoFeNiTi	AM	FCC	Ordered HCP	Tsau et al. (2009)
AlCrSiTiV	Laser cladding	BCC	(Ti,V)$_5$Si$_3$, Al$_8$(V,Cr)$_5$	Huang et al. (2011)
Al$_x$Co$_{1.5}$CrFeNiTi$_y$	AM	$x = 0$, $y = 0.5$: FCC	(Ni,Co)$_3$Ti	Chuang et al. (2011)
		$x = 0$, $y = 1$: FCC	(Ni,Co)$_3$Ti	
		$x = 0.2$, $y = 1$: FCC	(Ni,Co)$_3$Ti	
AlCoCuFeNiZr	SC	BCC + FCC	ZrFe$_3$Al	Zhuang et al. (2013)
AlCuMgMnZn	IM	HCP	Quasicrystal	Li et al. (2011)
(AlCuMnZn)$_{100-x}$Mg$_x$	AM	$x = 20$: HCP	Al−Mn quasicrystal	Li et al. (2010b)
		$x \neq 20$, HCP	Al−Mn quasicrystal, Mg, Mg$_7$Zn$_3$	

Table A2.6 Typical Strong HE Nitride and HE Carbide Films

HE Nitrides and Carbides	Hardness (GPa)	Young's Modulus (GPa)	Reference
(AlCrTaTiZr)N	36	360	Lai et al. (2006a)
(AlCrMoSiTi)N	35	325	Chang et al. (2008)
(AlCrSiTiV)N	31	300	Lin et al. (2007)
(AlBCrSiTi)N	25	260	Tsai et al. (2012)
(AlCrNbSiTiV)N	42	350	Huang and Yeh (2009)
(AlMoNbSiTaTiVZr)N	37	350	Tsai et al. (2008a)
(AlCrTaTiZr)C	40	303	Yeh et al. (2012)
(CrNbSiTiZr)C	33	360	Yeh et al. (2012)

Table A2.7 HEA Metallic Glasses

Composition	Processing Route	Reference
Equiatomic Substitution Alloys		
$(TiZrHf)_{60}(NiCu)_{40}$	IC	Ma et al. (2002)
$(TiZrHf)_{50}(NiCu)_{40}Al_{10}$	MA	Zhang et al. (2003)
$(TiZrHf)_{65}(NiCu)_{27.5}Al_{7.5}$	MeS	Kim et al. (2006c, 2007)
$(TiZrHf)_x(NiCu)_{90-x}Al_{10}$	MeS	Cantor et al. (2002), Zhang et al. (2003), and Kim et al. (2003b,c, 2007)
$(TiZrNb)_x(CuNi)_{90-x}Al_{10}$	MA	Zhang et al. (2007)
$(TiZrHf)_{100-x-y}(NiCu)_xAl_y$	MeS	Kim et al. (2003a, 2004)
$(TiZrHf)_x(NiCuAg)_{90-x}Al_{10}$	MeS	Cantor et al. (2002), Kim et al. (2003b), and Kim (2005)
$(TiZrHfNb)_{90-x}(NiCu)_xAl_{10}$	MeS	Kim et al. (2003b, 2006a,b)
$(TiZrHfNb)_{90-x}(NiCuAg)_xAl_{10}$	MeS	Kim et al. (2003b)
$(CuNiPdPt)_{80}P_{20}$	Flux water quenching	Takeuchi et al. (2011)
$(CuNiPd)_{60}(TiZr)_{40}$	MeS	Takeuchi et al. (2013b)
$(TiZr)_{40}(CuNi)_{40}Be_{20}$	AM	Ding and Yao (2013)
Equiatomic Alloys		
HfNbZr	Co sputtering	Nagase et al. (2012)
BeCoMgTi	MA	Chen et al. (2010b)
AlBFeNiSi	MA	Wang et al. (2014a)
AlCrNiSiTi	MaS	Chen et al. (2005a)
AlCrTaTiZr	MaS	Chang et al. (2011) and Hsueh et al. (2012)
AlDyErNiTb	IM	Gao et al. (2011)
BeCoMgTiZn	MA	Chen et al. (2010b)

(*Continued*)

Table A2.7 (Continued)

Composition	Processing Route	Reference
BiCoFeMnNi	Electro deposition	Yao et al. (2008)
CaMgSrYbZn	IM	Gao et al. (2011) and Li et al. (2013b)
NbSiTaTiZr	MaS	Yang et al. (2009) and Tsai et al. (2011)
AlBFeNbNiSi	MA	Wang et al. (2014a)
AlCrTaTiZrRu	MaS	Chang et al. (2011)
AlMoNbSiTaTiVZr	MaS	Tsai et al. (2008b)
Nonequiatomic Alloys		
$AlCoCu_{0.5}Ni$	MA	Chen et al. (2009d,e)
$AlCoCrCu_{0.5}Ni$	MA	
$AlCoCrCu_{0.5}FeNi$	MA	
$CaCu_{0.5}MgSrYbZn_{0.5}$	IM	Gao et al. (2011)
$CaSrYb(Li_{0.55}Mg_{0.45})Zn$	IM	
$AlCoCrCu_{0.5}FeNiSi$	SC	Chen et al. (2005d)
$AlCoCrCu_{0.5}FeNiSi$	AM	Chen et al. (2005b)
$AlCoCrCu_{0.5}FeNiTi$	MA	Chen et al. (2009d,e)
$AlCoCrFeCu_{0.5}MoNiTi$	MA	

AM, arc melting; IM, induction melting; IC, injection casting; SC, suction casting; BS, Bridgman solidification; MeS, melt spinning; LENS, laser engineered net shaping; MA, mechanical alloying; MaS, magnetron sputtering.

Printed and bound by CPI Group (UK) Ltd, Croydon, CR0 4YY

03/10/2024

01040423-0002